Pollution and Poverty

Shubhanku Kochar · Rananjayaa Singh
Editors

Pollution and Poverty

Environmental Casteism and Dalit Literature

Springer

Editors
Shubhanku Kochar
Guru Gobind Singh Indraprastha University
Delhi, Delhi, India

Rananjayaa Singh
Independent Researcher and
Freelance Copyeditor
Rohtak, Haryana, India

ISBN 978-981-95-6875-8 ISBN 978-981-95-6876-5 (eBook)
https://doi.org/10.1007/978-981-95-6876-5

This Springer imprint is published by the registered company Springer Nature Singapore Pte Ltd.
The registered company address is: 152 Beach Road, #21-01/04 Gateway East, Singapore 189721, Singapore

Foreword

I was born in the mid-1970s in the Himalayan state of Uttarakhand. Growing in the salubrious environment of Kumaon I may not have been subjected to pollution, but I do carry distinct memories of all-pervasive poverty and casteism. Amidst the picturesque Himalayas, also known as the water tower of Asia, a section of people was not allowed to drink water from the vessel meant for "savarnas". Though I didn't witness any physical violence against Dalits, but I could always sense a certain wretchedness, an ingrained cruelty displayed by the people of this prosperous mountainous land towards their fellow citizens, faint memories of which are still etched in my mind.

A young man sat on the ground, always carefully avoiding the cot or the chair nearby, and an elderly woman in her seventies touched the feet of a young bride. Such discriminatory customs, I later learnt, were not confined to my village but reflected the grim reality of a large part of India and the disparity eventually led to the unjust distribution of resources.

Be it a well or a grazing ground of cattle, public spaces were denied to some castes and none found it anomalous. The injustice based on birth was the immutable reality for many, the ugliness of which I could grasp only later.

Upon entering journalism, my work took me to different parts of the country and, thereon I began observing that the fault- lines were much deeper than I imagined. Conversations with marginalized communities made me realise that the sting was harsher and the social web more complex. For centuries the skewed distribution of ecological resources had engendered a carefully hatched system of injustice. True, a resistance had begun brewing over the years both at political and social level, but the situation had barely improved. While Dalits, at least to an extent, successfully found their voice in politics and decision making, adivasis, despite their peaceful protests (and sometimes armed resistance), largely remain neglected.

**

My earliest initiation into literature was through Hindi fiction. The tales of Renu, Premchand and Shivani had lent me a glimpse towards the injustice inflicted on Dalits, but the deep understanding of their crisis arrived only after I turned towards Dalit writers and thinkers. Dr B. R. Ambedkar and his works were in the curriculum, as teachers at school and elders at home often discussed the making of the great leader. While his greatness mostly remained limited to my young mind as a constitution maker, the real import of atrocities against Dalits and their struggle became clear after reading the life accounts of Dalit writers including Om Prakash Valmiki (Joothan), Sharan Kumar Limbale (Akkarmashi) and Tulsi Ram (Murdahia). The rebellion by Periyar, the reforms by Jyotiba and Savitri Phule expanded my horizon of understanding.

The bitter truths I witnessed on the ground as a journalist was supplemented by theoretical propositions of Dalit writers. Most recently, I made a short film on silicosis where most of the victims of this hazardous work were Dalits or adivasis. Similarly, while covering the impact of arsenicosis I found that it was even more difficult for Dalits to access clean water, as in some places they were denied the access to deeper borewells which provided less contaminated water.

**

At some places the situation has improved over the years but a lot is still to be done. I wish I could film this aspect of ecological injustice towards poor Dalits and Adivasis in wider elaboration and clarity. I must commend the promising academicians, most of them young, for preparing this anthology which explains the facets of ecology and environment with a caste perspective. The contributors have explored the thoughts and writings of Dalit writers through the lens of ecological injustice, and explained their texts in lucid prose. By incorporating Dalit voices, the anthology, thus, becomes an important contribution in the world of environmental literature and fills a void.

As we stare at the worst environmental disaster and equitable and just distribution of resources is the need of hour, this book will prove to be an engaging read for readers who want to venture into the less beaten path of ecology and environment.

New Delhi, India Hridayesh Joshi

Hridayesh Joshi is an award-winning journalist, writer, and photographer. During his career spanning more than 25 years he has reported on a wide range of issues including politics, agriculture, health, rural affairs, and internal security. He is the author of the book *Rage of the River—The Untold Story of the Kedarnath Disaster* (2013) published by Penguin India. He also wrote the novel *Lal Lakeer* (The Red Line) (2016) set in Bastar, published by Harper Collins India. In 2023, his Hindi translation of Bill Aitken's travelogue *Footloose in the Himalaya* was published as *Aitken ka Himalaya*. His recent work is the translation of Ramachandra Guha's literary memoirs published under the name *Samne Aate hi Nahin* (2025).

He writes and reports widely on current affairs, mainly on health, climate change, environment, and energy. He has also produced long form videos, reports, and documentaries besides writing stories for various newspapers and websites.

Contents

Contributors

Amar Chakrabortty Department of English, Government Arts and Commerce College, Sami, Patan, Gujarat, India

Pooja Duggal Department of English, Guru Nanak Dev University, Amritsar, India

Ritam Dutta Department of English, Jadavpur University, Jadavpur, India

Shubhadeep Jana Department of English, Presidency University, Kolkata, India

Mohd Amin Khan School of Humanities and Social Sciences, Indian Institute of Technology, Indore, India

Shubhanku Kochar University School of Humanities and Social Sciences, Guru Gobind Singh Indraprastha University, New Delhi, India

Dhanesh Mankulam Faculty of Arts and Humanities, CHRIST University, Bangalore, India

Monika School of Humanities and Social Sciences, Indian Institute of Technology, Indore, India;
Boston University, Boston, MA, USA

Umesh Patra Department of English, School of Humanities and Languages, Mahatma Gandhi Central University, Motihari, India

Pritee Sharma Mehta Family School of Sustainability, Indian Institute of Technology, Indore, India

Ashima Shrawan Shri Bhagwandas Adarsh Sanskrit Mahavidyalaya, Haridwar, India

Navdeep Singh M. R. Govt. College, Fazilka, Punjab, India

Rananjayaa Singh Trivium Education Services Pvt. Ltd, New Delhi, India

Krishnan Unni P Department of English, Deshbandhu College, University of Delhi, New Delhi, India

Chapter 1
Introduction

Shubhanku Kochar and Rananjayaa Singh

Abstract The global history of environmentalism, as Ramachandra Guha observes, has long been shaped by an elitist and distinctly European perspective. This dominant discourse often sidelines countries of the global South, dismissing them as merely poor and developing, and thus peripheral to the environmental narrative. In response to this imbalance, Guha introduced the concept of "Environmentalism of the Poor," foregrounding the reality that nations in the global South disproportionately shoulder the burdens of climate change. Building on this corrective, the present chapter argues that even within these southern nations, certain groups, marginalised and exploited by their own dominant communities, bear an unequal share of environmental burdens. These underprivileged groups are frequently excluded from mainstream national conversations about nature and environmental justice. A poignant example is found in India, where Dalits and lower-caste communities have historically been relegated to the peripheries, both socially and environmentally. Mukul Sharma notes that Indian environmental history is often "written from above," systematically erasing Dalit experiences and perspectives. Advancing this line of inquiry, this chapter introduces and explores the concept of Environmental Casteism. To illustrate and substantiate this argument, we turn to Indian literature, with a particular focus on short stories. We demonstrate how Dalits have been denied access to environmental benefits within the traditional Indian framework and are instead relegated to the margins, forced to grapple with dirt and deprivation. Furthermore, we argue that conventional environmental symbols, such as the village, commons, fire, water, and animals, take on markedly different meanings when viewed through the lens of Dalit experience.

Keywords Environmental casteism · Dalit literature · Indian short stories · Nature · Poverty · Pollution · Indian literature

---

S. Kochar
University School of Humanities and Social Sciences, Guru Gobind Singh Indraprastha University, New Delhi, India
e-mail: shubhankukochar@ipu.ac.in

R. Singh (✉)
Trivium Education Services Pvt. Ltd, New Delhi, India
e-mail: rananjayaasingh@gmail.com

S. Kochar and R. Singh (eds.), *Pollution and Poverty*,
https://doi.org/10.1007/978-981-95-6876-5_1

1.1 Background

In 2023, we had submitted a project on *Environmental Racism* with reference to African American fiction at our university. During our discussions, a recurring question surfaced: "Is there any Indian equivalent of this phenomenon?" Time and again, we were asked: "How does this idea of Environmental Racism help you in understanding your own neighbourhood?" and "Can you apply this concept to your immediate surroundings?" We were fully convinced that our research was fresh, relevant, and self-contained. We saw no need to relocate it to our local context. We believed it was more important to be specific and accurate about the society we were analysing, rather than drawing hasty parallels. As a result, we completed our project within the stipulated time, received appreciation, and, believing our job was done, looked forward to a period of rest before pursuing another endeavour. But this proved to be an illusion.

Within a few days of our submission, we stumbled upon a podcast, A J Longreads, by Al Jazeera. Al Jazeera Media Network is an independent news organisation pioneering in new and emerging storytelling techniques, like podcasts, that cater to a new generation (Al Jazeera 2023). The episode, titled "How death and despair haunt Pakistan's Christian minority," was about a sanitation worker who died on duty in Karachi, Pakistan. According to the report by journalist Michael Clark, "Masih went missing on 7 May 2023 after being swept away during a suspected flash surge while he was clearing a sewer. As is often the case with sanitation workers in Pakistan, Masih had to perform his work without safety equipment or… (PPE). His body was only recovered twelve days after he went missing" (End Water Poverty 2023).

One day, we decided to revisit the podcast. It lasted not more than half an hour; but by the time it ended, the seeds of this present project had been sown. All we had to do was water them, nurture them, and wait for the right time to harvest. The podcast in question realistically narrated the story of a 31-year-old Christian sanitation worker, Babar Masih, in Karachi. It highlighted how, in Pakistan, most of the dangerous and degrading work—work perceived as dirty—is relegated to religious minorities. Most of the Christian community in Karachi and Lahore, as the podcast informed us, can be found employed in mines and factories as labourers, and a considerable portion of them works in sanitation at the lowest levels. As mentioned above, these workers often labour without protective gear and go down into the sewers to clean them manually. Many of them either perish in the process, or get gravely injured. They do this because they are poor and have no other way to make ends meet (Al Jazeera 2023).

Since we have already explored a similar topic in much detail earlier, we were easily able to draw many parallels. We immediately connected the contents and issues of this episode to the condition of African American sanitation workers in the America of 1960s. In the words of the well-known American historian Carl Zimring, these "…sanitation workers had to handle all sorts of materials from tree limbs to broken glass to biological wastes that could infect, poison, or injure them…under temperatures regularly exceeding 90 degrees, often without shade or breaks to get water. Sanitation workers could be maimed at any time, and crippling injuries were common" (2015). The workers went on strike in 1968 after one of their colleagues died on the job. Their demands were simple but urgent: equal pay to that of White workers, and humane working conditions. Dr Martin Luther King Jr. had been enroute to support this movement when he was assassinated. The memory of reading about this ordeal came flooding back, and suddenly, something that had remained repressed within us throughout the previous project began to stir.

After pondering over this idea, we began to situate and recognise similar conditions in our own surroundings. The more we thought about it, the more we found ourselves at a crossroads. After much reflection, we arrived at a moment of epiphany. We concluded that in India, the closest equivalent of race is caste. Just as American society has been historically segregated along racial lines, Indian society has been deeply affected and divided along the lines of caste hierarchy. Similar to how African Americans have faced systemic racism since their arrival in the New World, Dalits in India have endured caste-based exclusion for centuries. The same way that White Americans have exploited and marginalised communities of colour socially, culturally, and historically, the upper castes in India too have relegated and excluded the people belonging to the lower castes. Despite the vast geographical, linguistic, cultural, and political distances between them, African Americans in the United States and Dalits in India share a common history of denial, and a parallel legacy of oppression and resistance. Both groups have been structurally deprived of dignity and opportunity. Crucially, both have also been disproportionately exposed to environmental hazards. For decades, their lack of political representation has allowed this to continue unchecked.

This is where concepts such as environmental racism and environmental casteism intersect—each rooted in a unique context, but sharing a common logic of environmental exclusion imposed upon marginalised communities.

1.2 Environmentalism of the Poor

Before we define our understanding of the term Environmental Casteism, it is important first to examine the idea of Environmentalism of the Poor, as our conceptual framework seeks to interlink both ideas. The history of environmentalism in its global context is highly contested and often one-sided. As Ramachandra Guha remarks in *Environmentalism: A Global History* (2000a), until recently, it appeared as though only the affluent nations of the First World had the luxury of championing "green" initiatives. The discourse often suggests that only the wealthy can afford to talk about and enjoy clean air and fresh water, while the poor, the oppressed, the crushed, the marginalised, the neglected, and the otherised, remain mute subjects, passive victims, not active agents of change.

Guha identifies three distinct waves of environmentalism:

1.2.1 The First Wave: The Romantic Reaction

According to Guha, the first wave of environmentalism began in the late eighteenth century and continued until the First World War in the West. This period saw a romantic reaction against the darker consequences of industrialisation. Writers and thinkers were appalled by the devastation wrought by the Industrial Revolution. Poets like William Wordsworth and thinkers like John Ruskin issued a call to return to the land. They idealised village life in stark contrast to the overcrowded, polluted urban centres springing up across Europe. This sentiment was mirrored across the Atlantic. In the United States, writers like Henry David Thoreau withdrew into the wilderness to rediscover harmony with nature. In such spaces, one could hear sparrows and nightingales rather than the roar of steam engines. One could see trees in abundance rather than industrial smoke and iron girders. In this context, the concepts of wilderness, national parks, and nature conservation began to gain prominence. However, these concepts were almost entirely designed *by* the rich *for* the rich. The "have-nots," the colonised, and the socially excluded were systematically dispossessed. As capitalism strengthened its grip on Western economies, commercialism and consumerism began to define human existence. The wealthy needed sanctuaries for weekend retreats: scenic mountains, verdant valleys, lush orchards, pristine beaches, and unpolluted waters. Having worked exhaustively throughout the week, the rich now sought to escape, if only for a while, to the "lap of Nature." For them, nature became a luxury, and a form of recreation.

1.2.2 The Second Wave: Protest and Policy

The second wave, according to Guha, unfolded during the latter half of the twentieth century, marked by a rise in environmental activism and a significant shift in civic consciousness. This shift began with the publication of Rachel Carson's *Silent Spring* (1962), which Guha aptly calls the "*Uncle Tom's Cabin*" of the environmental movement. With Carson's book, a profound transformation occurred: newspapers were filled with reports of ecological disasters, seminars and conferences proliferated, and civil society began to mobilise. Grassroots organisations and NGOs sprang up with a common purpose: to pressure politicians and citizens into saving the planet. This momentum was especially visible in America, where the first Earth Day in 1970 drew millions into the streets, and where environmental awareness became woven into the fabric of public life through events, rallies, and the institutionalisation of Environment Day (Sentinel Digital Desk 2024; Covering Climate Now 2020).

Importantly, this period also saw marginalised communities in the West, namely African Americans, Native Americans, and various immigrant groups, beginning to assert and demand their environmental rights. Their activism drew attention to the disparities in environmental quality: as Guha notes, affluent white neighbourhoods remained clean and well-resourced, while communities of colour were relegated to neglected environments. Zimring's *Clean and White* underscores this reality, describing how racial minorities endured overcrowded, unsanitary living conditions. These protests were fundamentally about survival—access to clean water, air, housing, and basic recognition. They were a powerful critique of the environmentalism that privileged the concerns of the elite.

Yet, it is critical not to converge these grassroots struggles of the poor in the West with what Guha and others later address as the "Environmentalism of the Poor." While marginalised communities in the United States and Europe were indeed gaining consciousness and mobilising against environmental injustices, their movements remained geographically and structurally confined to the Global North (Ciulla 2025). The emergence of environmental days, public symposiums, and widespread activism signalled a growing awareness, but this was still an environmentalism shaped by Western contexts and resources. These were, in essence, the struggles of the poor within wealthy nations. It was distinct from the Environmentalism of the Poor that would later arise in the Global South, where impoverished communities faced the compounded burden of environmental neglect, colonial histories, economic exploitation, and systemic exclusion from the global environmental discourse. Thus, while the second wave of environmentalism exposed the gaps and exclusions within elite-driven movements, it was only later that the voices and struggles of the Global South came to redefine the very meaning of environmentalism on a global scale.

1.2.3 The Third Wave: Environmentalism of the Poor

This leads us to the third and most urgent wave: *Environmentalism of the Poor*. Guha argues that this movement is deeply rooted in the Global South, where environmental struggle is not a matter of leisure or ideology, rather a matter of life and death.

To illustrate this, Guha offers five emblematic examples:

1. Penan Community, Malaysia: The Penan people of Sarawak are hunters and farmers who hold sacred their forests and rivers. In the 1980s, they faced off against commercial loggers who polluted their water, destroyed their forests, and endangered their lives. With the help of Swiss activist Bruno Manser, the Penan organised road blockades and successfully resisted corporate encroachment.
2. Medha Patkar and the Narmada Bachao Andolan, India: The proposed Sardar Sarovar Dam in Gujarat promised economic progress through irrigation and electricity. However, it threatened to submerge over 250 villages, destroy ancient temples, and erase dense forests. Activist Medha Patkar and her colleagues undertook massive awareness campaigns, fasting and walking through the Narmada Valley to expose the hidden cost of so-called development.
3. Thai Forest Farmers and Eucalyptus Monoculture: In the 1970s, the Thai state forest department proposed replacing diverse forest ecosystems with Australian eucalyptus trees to generate export revenue. This enraged local farmers who

depended on the forest for food, medicine, and fuel. Led by Buddhist monks, the villagers marched against the state's plans and forced a retraction.

4. Ken Saro-Wiwa and the Ogoni People, Nigeria: For 35 years, oil companies drilled the lands of the Ogoni people in Nigeria. The state benefited through revenue, but the people lived amidst devastation. There were no schools, no hospitals, no clean water, and no fresh air for the people. When social-rights activist Saro-Wiwa exposed this exploitation, the state executed him and eight of his colleagues in 1995.
5. Wangari Maathai and the Green Belt Movement, Kenya: Starting with just seven saplings in 1977, social and political activist Maathai mobilised thousands of rural women across Kenya to plant over seven million trees by 1992. Her work expanded the country's green cover and proved that ecological conservation could be rooted in local, communal action.

What all these struggles demonstrate is that Environmentalism of the Poor is not a theory debated in conference halls. It is not about curated treks, luxury eco-retreats, or beachfront cottages. It is a lived experience, and a grassroots struggle for survival. It is about:

- Farmers resisting the seizure of their fertile lands.
- Fisherfolk protesting dam projects that threaten their livelihoods.
- Forest dwellers safeguarding sacred spaces from corporate and state encroachment.
- Sanitation and mine workers demanding protective gear to shield themselves from toxic exposure.
- Rural communities marching towards national capitals to prevent their lands from becoming dumping grounds for industrial or nuclear waste.

In short, Environmentalism of the Poor is the struggle of the lower castes, the labouring classes, the marginalised, the ostracised, and the excluded. It is their assertion of the fundamental right to live—to breathe clean air, drink pure water, and occupy their rightful space in the ecology of the nation. This is not just a movement. It is life in action.

It is within this powerful context that we now turn to the concept of Environmental Casteism, which we see as the articulation of the Environmentalism of the Poor in the Indian context. Through close readings of Dalit literature, we explore how caste not only shapes social and political life, but also determines environmental access, interaction, and exclusion. Literature, in this case, becomes both a witness and a resistance.

1.3 Environmental Casteism and Literature

The Mahars' living conditions were wretched. In each little cubby-hole, there were three or four sub-tenants...The men worked as porters. Some went to work in mills and factories...The women...slaved even more than their men...[collecting] rags, papers, broken glass, iron and bottles in the street...sort them out all night...and sell them in the morning. (Pawar 2008, p. 10)

In our colony...
teachings get confused
civilization stumbles
The sun—even he is darkened. (Meshram 2009, p. 22)

In the lush green jungle
those tattered huts...
In that wild wood, wild people
eating roots and bulbs. (Sonawane 2011, p. 33)

Together, these verses evoke more than just deprivation; they speak not only of ecological deprivation but of a deeper symbolic erasure. They gesture towards a reality where even nature has turned its face away. The sun loses its light, roots become the only food, and civilization itself stumbles. In these lines, what's described is not only poverty or neglect, but a kind of deliberate environmental abandonment. Judith Misrahi-Barak offers an extension of this truth. Reflecting on the metaphors in Dalit testimonios, she recalls a moment where "stale crumbs smeared with dung and urine lying on the floor of an animal pen" come to stand not just for material degradation, but for a much deeper, symbolic violence—what she calls "a symbol of the outcasts' extermination... the annihilation of everything that makes the upper castes human" (Misrahi-Barak and Abraham 2018, p. 71). This isn't just about lack of access to nature. It's about being written out of its meaning. What nurtures others is what injures here. What liberates elsewhere becomes a cage.

The mainstream environmental discourse in India, as Mukul Sharma argues in *Caste and Nature: Dalits and Indian Environmental Politics* (2020), has largely been framed by dominant social groups and is often presented as a top-down narrative. Although India is a vast nation with immense ecological and cultural diversity,

Sharma insists that it is crucial to re-examine environmental history from below—from the ground of everyday caste experiences, particularly those of Dalits. These are the voices that have long been left out of mainstream narratives on nature.

According to Sharma, Dalits, like other marginalised communities, have always been active participants in shaping environmental history. Their day-to-day interactions with land, water, animals, forests, and soil constitute what Sharma calls a *cartography of environmental history from below*. To articulate the idea of Environmental Casteism more clearly, we now move through three interconnected domains within Dalit literature:

- The concept of the village
- The access (or lack thereof) to the commons
- The representation of animals

Through these lenses, we explore how caste fundamentally reconfigures ecological relationships, and how Dalit literature responds to and reclaims these relationships through its own aesthetic and ethical paradigms. We argue that categories of exclusion such as caste, race, ethnicity, and gender, while significant, are ultimately incidental to the broader question of environmental justice. At its core, there exist only two kinds of environmentalism: the environmentalism of the rich and the environmentalism of the poor. To support this claim, we draw upon examples from Indian short stories that depict both Dalit and non-Dalit lives. The stories have been selected at random to ensure objectivity in our analysis. We begin by examining how key environmental concepts are portrayed in non-Dalit contexts, and then demonstrate how their meanings shift when applied to Dalit realities.

1.3.1 Dalit Literature and the Village: Romanticism Versus Reality

The idea of village is often romanticised in Indian literary traditions. Sharma notes that India is widely seen as a land of farmers, and by extension, as a land of virtue and simplicity. The village is idealised as a golden realm where milk and honey flow, where fresh air and clean water are abundant, and where people live in harmony and innocence. This idyllic notion finds strong endorsement in Gandhian thought. Gandhi envisioned the village as self-sufficient, peaceful, hygienic, and spiritually sound—a place where humans coexist in harmony with animals and the environment. He imagined clean roads, spacious houses, accessible wells, and green pastures. It was, for Gandhi, a near-utopia, devoid of violence, treachery, and filth.

However, as Sharma forcefully argues, this image only applies to the upper castes. Traditional Indian villages were divided across rigid caste lines narrating twin histories of inclusion and exclusion. The people belonging to the upper-castes controlled access to the most fertile lands, fresh water sources, dense woods, rivers, and lush pastures. In contrast, the lower castes, especially Dalits, were confined to peripheries, with no access to clean drinking water, no entitlement to fertile soil, and no proximity to sacred groves or temples. O.V. Vijayan's Malayalam short story "The Wart" (1989) offers a stark illustration. The protagonist struggles relentlessly to rid himself of a persistent wart that keeps reappearing on his face despite undergoing multiple treatments. In the course of this ordeal, he loses both his wife and his only son—though these personal tragedies lie beyond the scope of our current analysis. What is of importance here is that the unnamed protagonist lives in a village where he is one of the most powerful landlords. For him, the village is nothing short of paradise. Vijayan offers vivid descriptions of the comfort and luxury that define the protagonist's life: "Ours was a country mansion, a granary fortress a hundred years old. Our estate had dwindled, but there was enough to ensure a life of leisure and contemplation. I did not know the extent of the lands nor how much grain they yielded, but Chaaththan, the head serf kept count of things for me as had his father for mine" (2001).

The above quotation reveals two contrasting realities. On one side stands the protagonist, for whom life is filled with ease and abundance. He owns a sprawling mansion and vast tracts of land, the extent of which he himself does not know. He need not work to make ends meet. His days begin before sunrise with a swim in the nearby river, followed by meditation and a leisurely breakfast. He spends the rest of the morning reading or wandering aimlessly. The afternoons are reserved for rest, while evenings are marked by tranquil walks with his eight-year-old son. Together, they visit the temple and sit under the moonlit sky, enjoying the sound of the waterfall. Their return is unhurried, and the day ends with a peaceful dinner. For the protagonist, the sun symbolises joy, the moon signifies comfort, and rain brings hope. On the other side is Chaaththan, the head serf, whose life is defined by labour and subservience. This stark contrast applies not only to Chaaththan but also to his ancestral lineage. Unlike the protagonist, Chaaththan must take orders, maintain records, and work tirelessly on land that does not belong to him. His existence is one of duty and discomfort, not leisure.

1.3.2 The Commons: Spaces of Inclusion or Instruments of Exclusion?

Mukul Sharma emphasises that in traditional Indian villages, *the commons*, rivers, wells, ponds, groves, and forests were regulated spaces, deeply shaped by caste. While commons are defined as shared resources meant to serve the entire community, dominant castes often maintained exclusive control over them. Derek Wall, in *The Commons in History*, affirms that the ideal of the commons has historically been subverted by dominant power structures. In "The Wart," we see the upper-caste protagonist enjoying unlimited access to the commons. He roams freely across riversides and hilltops, gathering medicinal leaves in whatever quantity he desires.

He dives into the river and swims for hours, undisturbed, because no one dares ask him to stop. His animals graze freely, asserting unchallenged dominance over the green pastures. He spends hours seated on the rocks near the village temple, gazing at the waterfall, immersed in the serenity of a world that bends entirely to his will. So complete is his sense of ownership that he comes to perceive even Chaaththan's young sister-in-law as part of the landscape he controls. When she shows him kindness during his isolation (after being abandoned by his wife and son) he violates that trust by sexually assaulting her, an act that ultimately leads to her death. Yet, the protagonist remains emotionally unmoved. His concern is not for the girl, but for the stubborn wart on his face, as though her life and death are of no consequence in his world.

In contrast, Avinash Dolas's "The Victim" (2001) presents the commons from the viewpoint of the Mahar community. Here, there is no utopia. There is only violence, exclusion, and fire. The story opens in a state of ruin: the narrator's mother is on the verge of death after surviving an arson attack. Her family has been waiting for her to die for the past five days. Their huts were burnt by the upper castes. With utmost difficulties, the unnamed narrator, a Mahar youth, was somehow able to pull his mother out of their burning hut. Their only "crime" was daring to contest the local panchayat election against an upper-class candidate. Their demand? Access to the panchayat well: a shared source of water, a common, that has always been denied to them. The consequences were devastating. As the narrator recounts: "One night our huts were set on fire and burnt down. Everything went to rack and ruin. Mother, encircled by flames, was dragged out from the hut. She could not be removed to the hospital and treated for her burns. Sitya, who went to call the police, was found dead in the river, with his backbone broken. All roads were blocked" (Dolas 2001).

For the upper castes, fire is a symbol of purification, and water represents nourishment. As Sharma observes, in Indian culture, water is revered as a cleanser, both physically and spiritually. Rivers such as the Ganges, Yamuna, Narmada, and Godavari are considered sacred; a dip in their waters is believed to absolve devotees of their sins. However, this symbolic reverence does not extend to the Dalit context. As Sharma further notes, people from lower castes have historically been denied access to common resources like rivers and wells on the basis of caste. This systemic exclusion has given rise to what can be termed *Environmental Casteism.*

When Dashrath Manjhi, the famed "mountain man" of India, as Sharma recounts, cuts through a mountain bit by bit, his act is more than one of resilience. It is a radical reclamation of space in the commons, a space that his community had been denied for centuries. In the short story under discussion, elements like fire and water, which are traditionally associated with cleansing and life, take on an opposite meaning for the Mahar community. Instead of being sources of purification, they symbolise exclusion, dispossession, and destruction.

When the narrator, his mother, and their community attempt to draw water from the Panchayat well, they are asserting a long-denied right to the commons. They are, in effect, demanding a share in the village's social and environmental resources. Tragically, this act of resistance invites violent backlash from the dominant castes. As the story concludes, the narrator reveals that, along with the death of his mother, their collective spirit to claim space and dignity also perishes. Thus, while the village may appear idyllic and abundant for the upper castes and landlords it is a site of exclusion and struggle for Dalits. This fundamentally alters the meaning of the "village" across caste lines.

1.3.3 Human-Animal Relations: Affection or Exploitation?

Another key site where Environmental Casteism unfolds is in the relationship between humans and animals. For dominant groups, animals are often figures of affection, spirituality, and connection with nature. In contrast, for Dalits, their relationship with animals has historically been shaped by degradation and survival. Ponkunnam Varkey's "The Talking Plough" exemplifies the former. The protagonist, Ouseph, is a Christian farmer whose bond with his bullock, Kannan, is deeply emotional. Kannan is not just a beast of burden; he is family: "Kannan was Ouseph's life. Grey in colour, sturdy in build...even Kannan's gait had a special aplomb" (Varkey 1959). Kannan is depicted in the story as a vibrant, sentient being, fully capable of expressing emotions. He responds intuitively to Ouseph's needs and displays a remarkable sensitivity, even to music—swaying gently whenever Ouseph sings. Kannan's affection is evident in the way he tenderly licks the sweat from Ouseph's body when the latter is exhausted, offering comfort and companionship. However, as the story nears its conclusion, Ouseph faces a heartbreaking dilemma. Pressured by the need to provide an adequate dowry for his daughter, he is forced to sell Kannan. The emotional weight of this decision is so overwhelming that Ouseph cannot bear to witness the transaction himself and deliberately stays away from the scene, underscoring the deep bond between man and animal and the pain of their separation. Later, when Ouseph discovers Kannan among the animals destined for slaughter to provide meat for the Easter feast, he is overcome with distress. Without hesitation, he spends all his remaining money to buy Kannan back. His wife and daughter are astonished by this decision, but Ouseph gently explains to his daughter that Kannan is as precious to him as she is. In the story's poignant conclusion, Kannan dies peacefully in his own barn, as if understanding that his passing might somehow protect his beloved master. The relationship between Ouseph and Kannan transcends the conventional dynamic of master and servant; it is more like that of father and son, with each defined in relation to the other. In this bond, neither the head of the family nor the animal he cherishes is seen as dirty or unworthy. Instead, their connection is one of mutual respect, love, and dignity.

Quite contrary to this, "Buffaloed" by Dalpat Chauhan offers an altogether different view. Set in Gujarat, it is the story of a Dalit couple, Ranachhod and Jivi. One day, Jivi is asked by the rich landlady of the village to take away a dead buffalo. In traditional Indian villages, Dalits or lower-caste people were expected to keep the village clean by dragging away the carcasses of dead animals. This was something they were forced to do in the name of tradition. However, on this particular day, when the story opens, it is an auspicious occasion, and as a matter of charity, the landlady decides to "donate" a dead buffalo from her household. When Ranachhod, Jivi's husband, is about to skin the animal, two things happen that disturb him. First, the carcass is attacked by dogs and vultures. At this, Ranachhod reflects that his life is utterly meaningless. He felt that it was a shameful matter that he has to fight with animals for food. Second, just as he is about to chop off the tail of the buffalo, it jumps and tries to run. Ranachhod is shocked to see that the buffalo was still alive. It then falls down with a thud and dies on the spot. Later, filled with guilt, Ranachhod keeps a day-long fast to purge himself of the sin of killing a living being. It turns out that the landlady had quietly told Jivi that the buffalo was still breathing, but was about to die, and could be taken away. Towards the end of the story, when Jivi serves Ranachhod a meal made from the same buffalo's meat, he feels as if the animal is rising from the food and roaming in his backyard. Overwhelmed by guilt and remorse, he decides to observe a fast on that very day.

As Mukul Sharma relates, Dalits have had to face animals on a daily basis in multiple contexts. Their lives were marked by constant confrontation with animals; sometimes as carriers of carcasses dragged out of the village, at other times as keepers of animals to sustain their age-old caste-based occupations. Often, they found themselves competing with animals for the same food. At other times, they survived by eating the meat of reptiles and insects, big or small. There was no dignity in their day-to-day interaction with animals. Unlike the non-Dalit experience, where animals could be companions or even sacred beings, there was no space for emotion in such dealings for Dalits. There was no friendliness, and no sense of connection. In most cases, Dalits had to manage dead animals, or navigate life around creatures considered dirty, or untouchable, like pigs, rats, and the like. For them, it was purely a matter of survival. It was not choice but caste-determined compulsion that shaped their response to the animal world. Denied the means to own cows or buffaloes for milk or livelihood by the powerful and the privileged, they were left to relate only with the discarded, the stinking, and the unwanted creatures of society.

Through these literary narratives, it becomes increasingly clear that caste operates not only at the level of politics, economics, or religion, but also manifests environmentally. It governs access to space, to commons, to animals, and to symbolic elements such as water and fire. Just as the field of Environmental Racism emerged from the need to examine "the racial disparities in environmental exposures and their adverse consequences" (Singh 2024), we must also recognise the caste-based environmental injustices within India. As we argued in our previous project, "the environment does not discriminate; it is society that discriminates" (Singh 2024). This is a claim that now takes on new gravity in the context of caste.

Environmental Casteism, thus, becomes a necessary term to describe how caste-based hierarchies are enacted and reinforced through unequal access to natural resources, environmental spaces, and ecological dignity. It is, in many ways, an Indian iteration of what American environmental sociologist David N. Pellow when he writes, "Racism extends its reach into all aspects of life, and the environment is no exception" (2004). This is a truth that holds just as urgently when caste replaces race in the Indian social matrix. Likewise, the assertion of another renowned environmental sociologist, Dorceta Taylor, finds clear echoes in the caste-based exclusions portrayed in Dalit narratives. She states, "the environment is a social product and, as such, is deeply influenced by societal power dynamics" (2014). In a similar vein, environmental historian Carolyn Merchant notes, "The environment is a theater in which social, economic, and political inequalities play out" (2007), and in the Indian context, caste decides who gets to play what role, or even enter the stage at all.

In the next section, we articulate the concept of Environmental Casteism as a critical intervention, building on the idea of Environmentalism of the Poor while rooting it within India's caste-based ecology and the lived experiences of Dalit communities.

1.4 What Is Environmental Casteism?

Drawing on the insights of Ramachandra Guha and Mukul Sharma, Environmental Casteism refers to the systematic discrimination faced by Dalits in their engagement with land, water, animals, and other natural resources—not just in political, social, or economic terms, but also environmentally. It is the articulation of a truth long known but rarely named: that caste defines access not only to rituals, temples, education, or public offices, but also to wells, forests, rivers, the commons, physical safety gear, and even clean air. Traditionally, Dalits were denied access to the commons, like rivers, ponds, groves, or village wells, because these were owned, regulated, or surveilled by upper-caste groups. This environmental exclusion was legitimised through religious purity codes and social hierarchies that branded Dalit presence as polluting.

Dalits were also denied the right to experience animals as companions or spiritual beings. Unlike the upper castes, who could own cows, bulls, or goats and treat them with affection, Dalits were restricted to carcasses, pigs, rats, and other beings socially coded as "dirty." Their interaction with animals was dictated not by empathy or livelihood alone, but by caste-assigned roles of waste disposal and street cleaning.

Water, which is considered a sacred purifier in the upper-caste imagination, often symbolises denial and violence for Dalit communities. The same well that brings life to an upper-caste household becomes a site of humiliation when a Dalit is beaten or killed for trying to drink from it. Similarly, fire, sacred in the context of Hindu rituals, becomes a tool of punishment when Dalit homes are set ablaze for asserting rights or breaching caste codes.

These exclusions did not end with the village. When Dalits migrated to the cities in search of better lives, they found themselves forced into caste-prescribed roles once again; this time as manual scavengers and sanitation workers. Even in modern urban spaces, the ecological burden remained the same; it was just clothed better in bureaucratic language. According to government data, "an overwhelming 97% of manual scavengers in India are Dalits, highlighting the strong link between caste and this practice…this data starkly illustrates the caste-based nature of manual scavenging and underscores the need to recognise and address this issue as a form of caste-based violence and discrimination" (Human Rights Watch 2014). Environmental Casteism is, therefore, not an isolated subcategory. It is a substructure of what Guha calls the

Environmentalism of the Poor. The key distinction between a marginal farmer in the Global South struggling to protect his land and a Dalit labourer in India lies in the depth of historical denial: the latter was not only dispossessed of land and resources but also systematically stripped of the very consciousness of environmental rights. For generations, Dalits were conditioned to believe that their birth destined them for the most degrading tasks, like handling excrement, sweeping waste, and living amid filth, because they themselves were deemed impure. For centuries, Dalits were denied the right to imagine clean water as a basic entitlement, green land as home, or a tree as companion. Their exclusion extended beyond the boundaries of mainstream society; it reached into the very fabric of nature. They were not merely dispossessed; they were *de-natured*. As the leading critic Judith Misrahi-Barak powerfully notes, "The hierarchical caste system traditionally implied a strict partitioning of space and physical segregation… Dalits were often relegated outside the village or at its margins… This physical apartheid corresponds to a symbolic expropriation of humanity itself" (Misrahi-Barak and Abraham 2018, p. 71). In a related vein, the renowned Dalit scholar Anand Teltumbde also asserts, "Dalits have been made to live in the filthiest conditions, denied the right to clean water, green surroundings, or even the shade of a tree, as if nature itself was forbidden to them. Their very touch, shadow, and voice were considered polluting, extending their exclusion beyond society to the environment itself" (Teltumbde 2013).

Today, however, the situation is not the same as it used to be during the pre-Ambedkar era. With education of the masses, constitutional safeguards, and grass-roots political assertion, people who were kept out of the mainstream society are defying cast hierarchies and demanding ecological justice. Yet, the historical trauma remains.

We now turn to the structure of this edited volume, outlining how each chapter contributes to the re-mapping of environmental discourse through Dalit perspectives. In the following section, we describe how each contribution interrogates, expands, and redefines the intersections of caste and environment, collectively advancing a more inclusive and critically engaged environmental scholarship.

1.5 Structure of the Book

This edited volume is structured into nine chapters, beginning with this chapter, this introductory essay, which outlines the theoretical framework, conceptual motivation, and methodological grounding of the project. The remaining eight chapters, written by contributing scholars, offer diverse but interlinked explorations of caste, ecology, and environmental marginality through literary, and interdisciplinary lenses. While each essay stands independently in its analysis, together they speak to a shared intellectual urgency: to reframe the discourse from the perspective of those historically excluded from both environment and narrative.

Chapter 2, titled "Silent Echoes: Dalit Women and Environmental Justice—A Literary Exploration" has been written by Monika, Mohd Amin Khan and Pritee

Sharma. It centres Dalit women's narratives and foregrounds the entangled operations of caste, patriarchy, and environmental degradation. Through the autobiographical voices of Baby Kamble and Bama, this chapter offers a textured analysis of how ecological deprivation, manifest in polluted water, toxic environments, and spatial exclusion, disproportionately burdens Dalit women. Yet these are not mere chronicles of suffering. The texts reclaim dignity and resistance, presenting ecological memory as a site of both oppression and agency. Kamble's fierce assertion of her Mahar identity, even amid brutal material deprivation, transforms her life-writing into a terrain of reclamation. Bama, similarly, critiques the environmental injustices woven into the religious and rural fabric of her community, deploying education, labour, and writing as acts of spatial and cultural resistance. Drawing on Mukul Sharma's caste-environmental paradigm and Malini Ranganathan's concept of "environment unfreedom," the chapter reconceptualises ecological harm as structurally entrenched within caste and patriarchy. Ultimately, it argues that environmental justice, when stripped of its caste-blind technocratic gloss, emerges as a deeply social and political imperative.

Chapter 3, "Voices of the Everyday: Dalit Aesthetics and Ecology in Select Short Fiction from Malayalam" by Dhanesh Mankulam, introduces the framework of Dalit aesthetics as a counter-discursive mode that challenges dominant, often sanitised, environmental narratives. It critiques the erasure of Dalit presence from mainstream literary ecologies by re-situating the land not as an idealised landscape but as a site of historical labour, pain, and remembrance. Focusing on two Malayalam short stories: T.K.C. Vaduthala's "Sweet-offering at Chankranthy" and M.K. Madhukumar's "Paalakkunnan's Journey," the chapter positions Dalit ecology within the everyday: embedded in ritual, emotional inheritance, and collective grief. These stories resist the rarefied aesthetic paradigms of both Sanskritic and colonial literary traditions, instead foregrounding nature as intimately interwoven with survival and resistance. Engaging thinkers such as Jacques Rancière, Michel Foucault, Alain Badiou, and Sharankumar Limbale, the chapter challenges the idea of aesthetics as apolitical. It extends the discourse of ecological casteism beyond material exclusion to encompass aesthetic deprivation—arguing that caste mediates not only access to physical resources, but also to beauty, representation, and the right to narrate one's ecological world.

Chapter 4, "Ecological Injustices and Dalit Narratives: A Reading of Hira Bansode's Select Poetry" by Ashima Shravan, turns to the powerful poetic corpus of Hira Bansode, one of the most incisive voices in Marathi Dalit literature, to explore the intersections of ecology, exclusion, and resistance. In poems such as "Petition," "Sanskriti," "O Great Man," and "Bosom Friend," Bansode conjures an environment not of natural abundance, but of betrayal and exclusion. Rivers remain unreachable, clouds withhold relief, and the land becomes a terrain marked by caste-inflicted denial. Through her spare yet potent verse, Bansode renders visible the quotidian realities of ecological injustice, where cooking with sawdust or being denied milk becomes an index of caste-based deprivation. Yet this is not merely an archive of suffering; it is a poetics of assertion. Her work resonates with dignity and pride in Dalit identity, forging a literary counter-space that defies erasure. The chapter draws on the work of Mukul Sharma, Gail Omvedt, and Gopal Guru to critically examine how dominant environmental discourse in India romanticises "traditional" ecological practices while remaining blind to their historical caste exclusivity. Bansode's poetry, in this context, operates as both testimony and resistance—insisting that environmental harm is lived, embodied, and political.

Chapter 5, "Roots of Exclusion: Examining Environmental Injustices in Balbir Madhopuri's *Against the Night* and Lal Singh Dil's *Poet of the Revolution*" by Navdeep Singh, shifts geographically to Punjab, engaging with the autobiographical and poetic voices of Balbir Madhopuri and Lal Singh Dil. It reveals how caste-based ecological violence persists even within regions celebrated for agrarian prosperity and social progressivism. Here, land, water, and even trees become sites of contested belonging. They are defined less by nature than by caste hierarchy. Madhopuri's and Dil's testimonies uncover spatial geographies of caste exile: Dalit homes pushed to the peripheries of villages, denied access to communal wells, punished for planting trees, and humiliated in both life and death. A banyan tree, once sheltering Dalit life, is cut down with symbolic cruelty; cremation grounds flood with sewage, marking even the ritual of death with ecological indignity. This chapter makes a crucial theoretical intervention by mapping environmental injustice not only onto material conditions but also onto symbolic and spiritual spaces. It argues that ecological violence operates as spatial apartheid, where even nature becomes implicated in the reproduction of caste. In doing so, the chapter reinforces the volume's core proposition: Environmental harm in India is never neutral, never merely environmental, but structured through the enduring hierarchies of caste.

Chapter 6, "Affective Land and Dystopian Environmental Desire in Raju K Vasu's *Polappatham*" by Krishnan Unni P, offers a philosophically layered reading of Raju K. Vasu's *Polappatham*, a novel that complicates the discourse of environmental injustice by foregrounding memory, alterity, and affect. Moving away from reductive binaries of access and denial, the chapter explores how Dalit subjectivity in Kerala, especially among the Pulaya and Paraya communities, navigates fractured cosmologies, internal caste hierarchies, and the residual legacies of colonial missionary interventions. Unlike narratives that romanticise indigenous ecological wisdom or rural harmony, *Polappatham* unsettles any such idealisation. Here, nature is neither benevolent nor redemptive; it is a site of tension, disorientation, and spiritual estrangement. The novel's engagement with land and ritual is deeply mediated by caste, rendering environmental affect as historically conditioned and politically charged. This chapter makes an especially significant contribution to Dalit ecocriticism by focusing on Kerala, a region often bypassed in pan-Indian discussions of caste ecology. It contends that ecological justice must extend beyond resource distribution to include ontological restoration—of memory, dignity, and identity. In this context, *Polappatham* becomes not just a literary text, but a philosophical intervention into the complex terrain of environmental casteism in South India.

Chapter 7, "Ecology of Plight: Reading of Selected Bangla Dalit Short Stories" by Umesh Patra and Shubhadeep Jana, turns to four Bangla short stories from *Survival and Other Stories: Bangla Dalit Fiction in Translation*, edited by Singha and Acharya, to investigate how caste articulates itself through ecologies of danger, deprivation, and survival. Here, the authors portray that caste violence does not always announce itself through overt slurs or ritual prohibitions. Instead, as they argue, it often hides behind economic precarity, environmental accident, and bodily vulnerability. In *"Survival"* by Shyamal Kumar Pramanik, a landless labourer fights a cobra for grain. His struggle dramatises both ecological desperation and casteist abandonment. *"On Water and On Shore"* by Gobinda Shoundo follows a fisherman who survives a sea storm only to be exploited by an upper-caste merchant on land, while *"Munnali"* by Manohar Mouli Biswas reveals the aspirations and degradations of a teenage scavenger caught between caste and city. In *"Farmer Gopal's Caste and Creed"* by Makhanlal Pradhan, a displaced farmer is coerced into urban anonymity, his caste trailing him like a silent wound. The authors have portrayed these stories to articulate what might be termed the "mutated ecology" of caste, where old exclusions take new forms, and survival itself becomes an environmental ordeal. Fiction here does not merely represent ecological injustice; it becomes ecological testimony, bearing witness to the evolving forms of untouchability in both rural and urban ecologies. This chapter thus deepens the volume's commitment to documenting the shifting terrains of caste and environment in contemporary India.

Chapter 8, "Stitched in Struggle: Dalit Clothing as Symbols of Environmental Struggle" by Ritam Dutta and Amar Chakrabortty, offers an original ecocritical intervention by reframing Dalit clothing as an environmental text. Drawing upon Roland Barthes, Dalit poetic traditions, and fashion theory, the chapter argues that garments are not merely cultural symbols but tactile repositories of caste-inflected ecological violence. Buttons missing from shirts, scavenged or discarded garments, ritual exclusions from certain fabrics; these are not incidental details, but material archives of pollution, labour, and systemic denial. Set against the backdrop of toxic labour zones, tanneries, waste sites, and industrial fringes, the chapter traces how the environments inhabited by Dalits leave imprints on clothing itself. These garments absorb more than dirt: they carry the invisible sediment of caste exclusion. Engaging deeply with thinkers such as Mukul Sharma, Sunita Narain, Rob Nixon, Sharmila Rege, and Kancha Ilaiah, the chapter critiques dominant environmentalism for neglecting this embodied materiality. The fabric of caste, it suggests, is quite literal—stitched into everyday dress and into the overlooked ecologies of survival. What emerges is a theory of "slow violence" encoded in cloth: persistent, intimate, and symbolic. Denied turbans, taxed saris, burned hems—these become ecological scars, worn in plain sight. Clothing, in this account, is both the medium and the message of environmental casteism, a second skin shaped by exclusion.

The final Chapter 9, "Caste, Deprivation, and Ecological Disparities: An Exploration of Environmental Injustice in *Outcaste: A Memoir*" by Pooja Duggal, returns to the autobiographical mode through Narendra Jadhav's *Outcaste: A Memoir*, placing ecological injustice at the heart of political resistance. Jadhav's recollections—of water deprivation, food scarcity, and caste-bound labour such as the Yeskar duty—reveal the intimate violence of ecological exclusion. These are not abstract resource concerns; they are daily humiliations lived through caste-marked bodies. Central to this chapter is the Chavdar Tank Satyagraha, recast not merely as a political milestone but as an ecological assertion. The act of demanding water, and of insisting on the right to drink, becomes an act of environmental reclamation. Food, water, labour, and dignity emerge here not as separate spheres but as entangled rights, all denied through the architecture of caste. The chapter thus closes the volume with an urgent provocation: that ecological justice in India can neither begin nor endure without the annihilation of caste. Jadhav's memoir is not just testimony; it is a political roadmap, reminding readers that true environmental reform must dismantle the very social hierarchies that structure ecological access.

Across these nine chapters, the volume insists on a redefinition of environmental justice; one that acknowledges caste not as an external social variable, but as an organising principle of ecological life in India. Three critical insights emerge:

1. Environmental deprivation is caste-determined, not merely developmental or economic.
2. Dalit literature redefines environmentalism by refusing to treat nature as a passive backdrop. It politicises space, reclaims labour, and documents survival as resistance.
3. Ecological futures cannot be caste-blind; any such imagination is structurally complicit in the exclusions it seeks to remedy.

Dalit literature, in this context, emerges as both archive and provocation: an archive of lived ecological memory, and a provocation to reimagine justice beyond technocratic solutions or romanticised indigeneity.

This volume thus calls on ecocritics, policymakers, and scholars to listen to voices from the margins, not as supplementary, but as foundational. In language, land, labour, and literature, caste persists as a deep ecological structure. To ignore it is not neutrality; it is complicity.

1.6 Limitations and Possibilities

While this volume offers a vital intervention into the intersections of caste and ecology, it is important to acknowledge its limitations with candour. The project is situated within the Indian context, yet the vast heterogeneity of India's ecological regions, caste geographies, and literary traditions means that any such selection will inevitably be partial. The focus here falls primarily on literary and autobiographical texts, privileging genres such as fiction, poetry, and memoir. In doing so, it does not sufficiently engage with oral histories, visual cultures, or digital storytelling—forms that have become increasingly significant in contemporary Dalit expression and ecological testimony. These alternative genres warrant deeper scholarly attention.

Regionally and linguistically, the volume draws from key literary traditions, namely Marathi, Malayalam, Bangla, Tamil, and Punjabi, but does not encompass the full cultural breadth of Dalits across India. Voices from the North-East, Central India, or Adivasi-Dalit intersections are, for instance, largely absent. This gap reflects a structural constraint, one that future research must strive to address with greater collaborative effort.

Methodologically, the volume is anchored in literary and ecocritical analysis, with engagements from philosophy, sociology, and history. Yet the urgent need for interdisciplinary synthesis, especially between the environmental humanities, policy studies, and grassroots activism, still remains. Only such cross-sectoral dialogue can address a more comprehensive vision of ecological justice in the India as we know it.

Ultimately, acknowledging these limitations is a recognition of the complexity of the field. The conversations initiated here are not conclusive. They invite further research that continues to centre marginalised voices while expanding the scope of the intersection of Dalits and the environment.

References

Al Jazeera. 2023. How death and despair haunt Pakistan's Christian minority (AJ Longreads Podcast, 14 October 2023). https://www.aljazeera.com/podcasts/2023/10/14/how-death-and-despair-haunt-pakistans-christian-minority

Alter, Stephen and Wimal Dissanayake (eds.). 2001. *The Penguin Book of Modern Indian Short Stories*. Penguin.

Chakrabarti, Ranjan (ed.) 2020. *Critical Themes in Environmental History of India*. Sage.

Chauhan, Dalpat. 2023. *Fear and Other Stories* (H. Ashwinkumar, Trans.). Penguin Random House India.

Ciulla, Alice. 2025. Creating the Global Environment from Below: The United States Environmental Justice Movement in the World Arena. *Global Environment* 1–29. https://doi.org/10.3828/whpge.63837646622521.

Covering Climate Now. (2020, April 22). The Radicalism of the First Earth Day. https://coveringclimatenow.org/from-us-story/the-radicalism-of-the-first-earth-day/

Crosby, Alfred. 1986. *Ecological Imperialism: The Biological Expansion of Europe*. Cambridge University Press.

Dangle, Arjun (ed.). 2009. *Poisoned Bread: Translations from Modern Marathi Dalit Literature*. Orient Blackswan.

Dolas, Avinash. 2001. "The Victim." In *The Penguin Book of Modern Indian Short Stories,* edited by Stephen Alter and Wimal Dissanayake, 279–85. Penguin Books India Pvt. Ltd.

End Water Poverty. 2023. #JusticeForBabarMasih: Death of Pakistani Sanitation Worker Exposes Inhumane Working Conditions.https://endwaterpoverty.org/justiceforbabarmasih-death-of-pakistani-sanitation-worker-exposes-inhumane-working-conditions/

Gadgil, Madhav and Ramachandra Guha. 2020. *This Fissured Land: An Ecological History of India*. Oxford University Press.

Guha, Ramachandra. 2000a. *Environmentalism: A Global History*. Longman.

Guha, Ramachandra. 2000b. *The Unquiet Woods: Ecological Change and Peasant Resistance in the Himalayas*. Expanded Edition. Oxford University Press.

Human Rights Watch. 2014. Cleaning Human Waste: 'Manual Scavenging,' Caste, and Discrimination in India. Human Rights Watch. https://www.hrw.org/report/2014/08/25/cleaning-human-waste/manual-scavenging-caste-and-discrimination-india.

Marzec, Robert. 2007. *An Ecological and Postcolonial Study of Literature from Daniel Defoe to Salman Rushdie*. Palgrave Macmillan.

Merchant, Carolyn. 2007. *American Environmental History: An Introduction*. Columbia University Press.

Meshram, Kishor. 2009. In Our Colony. In *Poisoned Bread: Translations from Modern Marathi Dalit Literature*, ed. Dangle, Arjun, 22. Orient Blackswan.

Misrahi-Barak, J., and J. K. Abraham (eds.). 2018. *Dalit Literatures in India* (2nd ed.). Routledge.
Pawar, Daya. 2009. Son, Eat Your Fill. In *Poisoned Bread: Translations from Modern Marathi Dalit Literature*, ed. Dangle, Arjun, 10. Orient Blackswan.
Pawar, Urmila. 2008. Motherwit. In *The Exercise of Freedom: An Introduction to Dalit Writing*, eds. K. Satyanarayana and Susie Tharu, 3–17. Navayana.
Pellow, David N. 2004. *Garbage Wars: The Struggle for Environmental Justice in Chicago*. MIT Press.
Sentinel Digital Desk. 2024. Environment Day 2024: Know the date, history and significance. *The Sentinel Assam*, May 31. https://www.sentinelassam.com/more-news/life/environment-day-2024-know-the-date-history-and-significance
Sharma, Mukul. 2020. *Caste and Nature: Dalits and Indian Environmental Politics*. Oxford University Press.
Singh, Rananjayaa. 2024. Environmental Racism: A Reading of Zora Neale Hurston's and Richard Wright's Selected Fiction. PhD Thesis, Guru Gobind Singh Indraprastha University, Shodhganga. http://hdl.handle.net/10603/585091
Sonawane, Shantabai. 2011. In the Lush Green Jungle. In *No Alphabet in Sight: New Dalit Writing from South India*, eds. K. Satyanarayana, and Susie Tharu, 33. Penguin.
Taylor, Dorceta. 2014. *Toxic Communities: Environmental Racism, Industrial Pollution, and Residential Mobility*. New York University Press.
Teltumbde, Anand. 2013. Dalits: Past, Present and Future. In *The Oxford India Anthology of Dalit Literature*, eds. K. Satyanarayana and Susie Tharu, 123–135. Oxford University Press.
Varkey, P. 1959. The Talking Plough. In *Contemporary Indian Short Stories: Series 1*, 45–52. Sahitya Akademi.
Vijayan, O. V. 2001. The Wart. In *The Penguin Book of Modern Indian Short Stories*, eds. Stephen Alter and Wimal Dissanayake, 99–109. Penguin.
Wall, Derek. 2014. *The Commons in History: Culture, Conflict and Ecology*. MIT Press.
Westling, Louise (ed.). 2014. *The Cambridge Companion to Literature and the Environment*. Cambridge University Press.
Zimring, Carl A. 2015. *Clean and White: A History of Environmental Racism in the United States*. New York University Press.

Shubhanku Kochar is an Associate Professor at University School of Humanities and Social Sciences (USHSS) at Guru Gobind Singh Indraprastha University, Delhi. He specialises in African Literatures in English, Eco-Criticism, and Literature and environment. His latest publications are *Environmental Post-Colonialism: A Literary Response* (2021, Lexington Books), *Literature from the Peripheries: Refrigerated Culture and Pluralism* (2023, Lexington Books), and *Pastoral and Anti-Pastoral: Representation of City and Village in Literature* (2024, Ibidem/Columbia University Press). Presently, he is working on a project with Springer, titled *Literary Representation of the Indian Migration during the British Empire*. His poems have been published in Protagonist, a famous literary magazine.

Rananjayaa Singh is a Tutor with Trivium Education Services, India, in collaboration with Tutor.com (Princeton, U.S.A.), since November 2024. She holds a PhD in English from Guru Gobind Singh Indraprastha University, New Delhi. She is an independent researcher and a freelance copyeditor and proofreader, helping students across the world refine their academic voices. Her research interests include literature and the environment, African American literature, and narratives from the margins. She serves on the editorial boards of Pacific Journal of Modern Theories and Research (PJMTR) (E-ISSN: 3108-3110) and The Criterion: An International Journal in English (ISSN: 0976-8165).

Chapter 2
Silent Echoes: Dalit Women and Environmental Justice—A Literary Exploration

Monika, Mohd Amin Khan, and Pritee Sharma

Abstract Dalits, historically known as the outcaste in India's hierarchical social order, endure systemic discrimination that intersects with gendered oppression, subjecting Dalit women to compounded vulnerabilities. Their lived experiences—shaped by caste and gender dynamics—remain strikingly absent from mainstream environmental discourse, despite disproportionate exposure to ecological degradation and resource deprivation. Using intersectional feminist frameworks, this chapter analyses the nexus of caste, gender and environmental justice (EJ) through the autobiographies of two seminal Dalit writers: Baby Kamble's *The Prisons We Broke* (1986/2008) and Bama's *Karukku* (1992). Kamble and Bama demonstrate how caste-based spatial segregation confines Dalit communities to environmentally hazardous peripheries, systematically denying access to clean water, arable land and sustainable livelihoods. Through close textual analysis, this chapter reveals how hegemonic power structures distribute environmental resources along caste lines, enforcing Dalits' relegation to toxic scarcity. Emphasising on Dalit feminist perspectives, the chapter critiques homogenised EJ frameworks, advocating for intersectional policy approaches to develop inclusive and sustainable solutions.

Keywords Dalit women · Autobiographies · Environmental justice · Equitable approach · Literature perspective

Monika (✉) · M. A. Khan
School of Humanities and Social Sciences, Indian Institute of Technology, Indore, India
e-mail: phd2101161006@iiti.ac.in

Monika
Boston University, Boston, MA, USA

P. Sharma
Mehta Family School of Sustainability, Indian Institute of Technology, Indore, India

S. Kochar and R. Singh (eds.), *Pollution and Poverty*,
https://doi.org/10.1007/978-981-95-6876-5_2

2.1 Introduction

Environmental Justice (EJ) can be defined both as a group of social movements and a body of research aimed at promoting and actualising distributive, procedural, and recognitional forms of justice (Johnson et al. 63). These forms of justice include fairness in the distribution of environmental benefits and harms, respect for and acknowledgment of alternative environmental worldviews, and equitable access to environmental decision-making. The U.S. Environmental Protection Agency defines EJ as "the fair treatment and meaningful involvement of all people, regardless of race, color, national origin, or income, with respect to the development, implementation, and enforcement of environmental laws, regulations, and policies" (US EPA 2024). The goal of EJ emphasises (1) the same degree of protection from environmental health hazards and (2) equal access to the decision-making process, ensuring a healthy environment to live, work, and learn. EJ advocates challenging the current environmental protection system and proposes alternative frameworks to address environmental injustices, disparate impacts, and unequal protection (Bullard 2001).

In India, Dalits—formerly referred to as "outcastes" and belonging to the lowest strata of society—face severe discrimination, depriving them of fundamental rights and environmental justice (Bhimraj 2020). They are relegated to peripheral areas on the outskirts of settlements, which are frequently repurposed as dumping grounds for pollutants and industrial waste, subjecting them to severe environmental hazards (Prasad 99). Despite efforts to promote social reform, they continue to face institutional barriers, limiting their access to resources and opportunities. This discrimination, deeply entrenched in societal structures, reinforces exclusion, specifically for women of Dalit community who bear the dual burden of caste and gender oppression (Pasic 2020). Women of these communities have to live with double discrimination in accessing environmental justice—first due to systemic caste oppression and second due to patriarchal norms, which push them into marginalised environments and perpetuate cycles of exclusion (The Dalit: Born into a Life of Discrimination and Stigma 2021).

Dalit literature, alongside Dalit feminist voices, has emerged as a significant form of resistance against entrenched caste prejudice (Oza 2020). Through their literary contributions, Dalit writers in particular have played an instrumental role in deconstructing hierarchical structures, promoting social justice, and envisioning a society that is more inclusive and equitable (Simon 2021).

Although, in recent years, the importance of intersectionality in understanding environmental issues, particularly within marginalised communities, has surged (Sharma 2019). Yet, the specific experiences of Dalit women in their interactions with the environment have largely been overlooked in both mainstream environmental discourse and Dalit literature. This gap underlines the need for a more in-depth investigation into how intersectional identities impact an individuals' interactions with nature. In this regard, Dalit literature, especially autobiographies of Dalit women, like Bama and Kamble, has become a vital form of resistance against entrenched systems of oppression (Rege 2014). Bama's *Karukku* and Kamble's *The*

Prisons We Broke offer first-hand accounts of the lived experiences of Dalit women, providing a critical lens to understand the intersection of caste, gender, and environmental injustice. Both authors offer profound insights into the compounded challenges faced by Dalit women, who are marginalised not only through caste but also through environmental exploitation.

Bama, through her autobiography, explores the emotional and spiritual struggles of Dalit women within a caste-based Christian community, highlighting how environmental neglect is entangled with social discrimination. Kamble, in her text takes a more socio-political approach, critiquing the institutional overlook of Dalit communities, particularly in rural areas, where caste-based environmental injustices are pervasive. While Bama focuses on the sensitive toll of caste and gender oppression, Kamble addresses the material and structural aspects of these issues. Both the authors, however, challenge dominant narratives and offer critical perspectives on the systemic exploitation faced by Dalit women. Their autobiographies are crucial to understanding the broader discourse of EJ, particularly in terms of how marginalised communities interact with their environment. The interplay of gender and caste relations creates inevitable obstacles, often rendering Dalit women's issues invisible in mainstream discourse. The continuing legacy of caste prejudice denies them essential rights, opportunities, and resources, pushing them to the margins of the society (Sharma 2012). Thus, through the exploration of Kamble's and Bama's autobiographies, this chapter investigates two major questions:

(a) How does environmental exploitation relate to caste-based discrimination, perpetuating cycles of poverty and deprivation among Dalit communities?
(b) How do caste dynamics influence resource allocation patterns, contributing to the marginalisation and deprivation experienced by Dalit communities in terms of access to and sustainability of the environment?

To critically engage with these questions, this chapter undertakes a close analysis of two seminal Dalit autobiographies—Bama's *Karukku* (1992), a groundbreaking late twentieth-century text, and Baby Kamble's *The Prisons We Broke* (originally *Jina Amucha*, 1986, translated into English by Urmila Pawar in 2008)—examining their narratives through the dual lenses of contemporary human dynamics and environmental injustice. By situating these works in dialogue with both foundational late twentieth-century Dalit literature and emerging post-2010 scholarship on caste and environmental justice, this study illuminates how systemic discrimination has historically shaped—and continues to influence—ecological marginalisation. Furthermore, it interrogates the evolving discourse on resource allocation, arguing for an inclusive reimagining of environmental access that positions Dalit lived experiences as critical to sustainable social transformation.

2.2 Literature Review: Contextualising Caste, Gender, and Environment

In India, the caste system is a deeply embedded social hierarchy that divides members of society according to their place of birth, occupation, and social status (Deshpande et al. 2016). Traditionally, this system has created strict Varna orders in Indian culture, with Brahmins (priests and scholars) at the top and Kshatriyas (rulers and warriors), Vaishyas (merchants and traders), and Shudras (labourers and artisans) at the bottom. The Dalits, formerly known as untouchables/outcaste, are a community of people who fall out these social orders (Bhattacharjee, 2023). Historically, they were seen as being outside the caste system and faced harsh social, economic, and religious discrimination. The caste system has been in place for a millennium, impacting social relationships, marriage, and employment, among other spheres of life (Rao 2014). However, caste-based biases and inequities still exist in contemporary Indian society despite legal abolition through the introduction of *Right to Equality* act as a fundamental right in constitution along with affirmative actions, such as reservation policies, socio-economic upliftment programs, and anti-discrimination laws (Srinivas 1957).

This hierarchical system controls access to resources, opportunities, and social status, maintaining long-standing prejudice and marginalisation. According to B.R. Ambedkar (1946), a towering figure in Indian history, social reformer, jurist, economist, and the principal architect of the Indian Constitution, the caste system is rooted in graded inequality, with Dalits experiencing significant social, economic, and political downgradation. The term *Dalit* in itself derived from the Sanskrit word *"dalan"* or the Hindi word *"dalita"* meaning "oppressed" or "broken." This exclusion is exacerbated by the interaction of caste and gender dynamics, which primarily affects Dalit women.

Dalit women experience numerous layers of biases due to their lower caste and gender, positioned at the critical juncture of caste, class, and gender inequalities, making them more vulnerable to exploitation and violence. In relation to EJ, Dalit women are disproportionately disadvantaged by environmental degradation and resource scarcity. Dalit women also carry the burden of obtaining and managing scarce environmental resources such as water, fuelwood, and agricultural land, all while suffering prejudice and violence from dominant caste groups (Dastidar 2021).

According to Mukul Sharma, a prominent environmentalist, numerous Dalit movements have emerged over time to address the link between caste, nature, and the unequal distribution of natural resources, viewing this connection as a crucial first step in combating environmental injustices (Sharma 2019). He also emphasises that Dalit livelihood issues and rights have gained prominence over the past 20 years in relation to environmental degradation, pollution, raw material extraction, ecosystem changes, pricing and marketing of resources, the effects of climate change, and the creation of environmental policy. Despite their growing access and participation in environmental movements in recent times, Dalit estrangement is very deep, and still, they remain detached from mainstream movements. This implies that an understanding of the existential, experiential, spatial, and cultural blinders—such as soil, water, air, touch, taste, smell, and space—is required in order to establish Dalit environmental justice and ecology (Sharma 2022a, b).

Environmentalists Srinivasan et al. (2022) underline how disproportionately the climate catastrophe affects underprivileged groups, stressing that Dalit women especially are vulnerable to environmental discrimination because of ingrained caste and gender inequality. At the intersection of caste and gender oppression, Dalit women endure systemic discrimination, which exacerbates their exposure to environmental hazards—including toxic air and water pollution, climate disaster-induced displacement, and livelihood loss. Yet, their struggles are often neglected in national discourse, further suppressing their voices. A political ecologist and geographer, Ranganathan (2021) employs the term "Environmental Unfreedom" to analyse the struggles of Dalit women in India, where caste, gender, and socioeconomic inequalities intersect. She argues that caste structures perpetuate unfreedom through the systemic segregation, containment, and disciplining of Dalit labourers—a process reinforced by the racialisation of property and personhood. Dalit women, as a result, bear the brunt of industrial labour exploitation and sanitation work. Additionally, caste plays a significant role in housing segregation and climate vulnerability, highlighting the need to consider caste as a central factor in understanding the spatial, labour, and ecological disparities faced by Dalits in urban India (Ranganathan 2022).

Furthermore, Behl and Kashwan (2022)—researchers specialising in food/water access and environmental studies, respectively—highlight the intersectional discrimination Dalit women face in accessing water, exacerbated by caste and gender inequalities. They point out that during Chennai water crisis in 2019, Dalit communities, especially women, were denied access to public water sources due to caste-based segregation. This crisis exposed how caste geography determines water access—with 78% of Dalit settlements in peripheral areas lacking piped water (TNUEF survey 2019), forcing women to depend on hostile upper-caste controlled sources. The scholars further contend that addressing water insecurity in India requires a multi-pronged approach that considers both neoliberal infrastructural issues and the persistent social structures of caste and gender oppression. They stress that traditional water governance systems, like water distribution committees, often fail to challenge caste and gender norms, reinforcing existing inequalities (Behl & Kashwan 2022).

Despite the growing scholarship on EJ, the specific experiences of Dalit women remain underrepresented in environmental, feminist, and literary studies. Existing literature in India often prioritises regional or class disparities, neglecting the intersection of caste and gender in EJ. EJ studies focus on urban–rural divides or industrial pollution but rarely address how caste-based segregation forces Dalit women into hazardous spaces with limited access to essential resources like fresh air, clean water or safe housing. Similarly, Dalit literature, though rich in socio-political critique, has historically overlooked environmental concerns associated with patriarchal challenges, framing them as secondary to caste-based oppression. This gap ignores how environmental degradation disproportionately impacts Dalit women, who bear the dual burden of caste and gender-based exclusion. Addressing this requires intersectional approaches that recognise these overlapping injustices. Dalit women's autobiographies, such as those by Bama and Kamble, challenge this exclusion, offering critical insights into the entwined nature of caste, gender, and environmental justice in India.

2.3 Environment Interaction Through Autobiographical Exploration

The emergence of Dalit and ecological writing enabled sociologists and anthropologists to examine the lived experiences shared by individuals from the lower strata of society (Naudet 2008). A more nuanced examination reveals how environmental casteism differentially impacts population groups—including men, women, and children—across intersecting axes of oppression. Baby Kamble's autobiography, *Jina Amacha,* which was initially written in Marathi, is a moving account of her life as a Dalit woman in India. The memoir, translated into English as *The Prisons We Broke,* offers engaging insights into Kamble's hardships and successes on her journey from oppression to empowerment. Born into a family of Mahars, who are considered untouchables in the caste hierarchy, Kamble clearly illustrates the brutal reality of caste-based prejudgment, poverty, and social marginalisation she experiences from a young age. Kamble's narrative reveals the various layers of systematic injustice inherent in Indian society, as well as Dalit communities' strength in resisting persecution. Like Kamble, Bama's phenomenal work *Karukku* narrates the same oppression faced by Dalit women in Tamil Nadu from a different perspective. It focuses on the key issue of caste oppression in the Catholic Church, illustrating the conflict between the individual and the community and depicting Bama's life as a path of introspection and healing from institutional and societal rejection (Naik 2016). These autobiographies provide poignant accounts of the discrimination they faced illuminating the challenges encountered by the authors, who were subjected to intersecting forms of oppression—initially as victims of patriarchal norms and subsequently as Dalit women.

These autobiographies enable us to better understand their historical and contemporary problems as Dalits interact with the environment, highlighting how the interplay between caste and gender differs from the broader impact of caste on the community as a whole. In her book, Kamble proudly states that she wrote her autobiography because she was proud to be born a Mahar, rather than feeling devalued by it. She mentioned that she was born in the region of Maharashtra, and the land itself bears their name, questioning, "Whom should they be lowered off…? We are the great race of the Mahars of Maharashtra. We are its real original inhabitants, the sons of the soil. The name of this land is also derived from our name" (Kamble 2017).

Kamble's narration expresses deep love for her community without romanticising their oppression. This fusion of compassion for her people's suffering and her anger at their inhumane conditions creates a tone that is both wise and self-reflective. The narrative never descends into self-pity. The autobiography, on the other hand, is subtly humorous, as shown by her descriptions of ritual bathing, dirty kids with runny noses, games, and weddings. Kamble depicts a world defined by 'difference in location'. The disparity is not limited to location. Her realm is situated on the outskirts of both the village and the social imaginary. It is perceived as unclean, impure, contaminating, and untouchable, creating a sense of alienation. In addition, they reflect Brahminical rule through superstitions, illiteracy, ignorance, and oppression. Talking about the prevalent patriarchal conditions she gives a vivid account of the psychological and physical abuse that women experience in both public and private settings. She mentions if Mahar women are the 'other' for Mahar men, then the Mahar community is the "other" for the Brahmins (Kamble 2017). Kamble shows us how patriarchy and caste work together to keep women exploited in society and emphasis that women's need to identify who they are is most apparent at this point. As a result of their struggles, the Mahar women have emerged as the community's agents of reform, and Kamble demonstrates their incredible dignity and resiliency.

Kamble's autobiography recounts numerous incidents that reveal how their life was an unending cycle of oppression. Yet, despite this, they never missed an opportunity to appreciate the environment they lived in, bowing to their motherland as true lovers and natives of the land. Talking about her childhood plays and nearby atmosphere Kamble wrote for them, even a cactus plant outside their home was a boon (Kamble 2017). They played with cacti, crafting them into ornaments—wearing thorny stems as nose rings—and even relied on them as a food source during hunger. This reveals how the Dalits, despite the injustice of their environment, never relinquished their reverence for nature. Kamble reclaims ecological belonging by depicting Dalits as earth's natural children who endure environmental hardships (thorns, heat) yet maintain sacred soil rituals until final reunification with the land (Kamble 2017). This narrative contests casteist notions of pollution by asserting an unbroken cycle of sustenance and return.

Kamble's biography frames environmental casteism through an interspecies lens, revealing how caste oppression dehumanises Dalits, relegating them to animal-like survival (Vishwakarma 2022). Her story exposes the dehumanising impacts of caste oppression, as she reveals how her tribe is treated like animals, lacking agency and dignity. It emphasises how higher-caste people methodically destroy their capacity for thought and thinking, bringing them to a state much worse than that of beasts. While animals brought up by higher castes are given sufficient food, Dalits are forced to rely on leftovers and live in the squalor of waste pits outside the hamlet (Kamble 2017). Their lives are marked by constant struggle as they strive for their meagre survival while navigating various natural disasters coupled with the humiliation of slavery. Kamble reconceptualises Dalit marginalisation as latent power: like the ocean absorbing sins without protest, their silent endurance structurally sustains caste society. This ecological metaphor subverts victimhood narratives, positioning Dalits as moral anchors deserving reverence (Kamble 2017).

Her story draws our attention to the prevalence of caste-based discrimination and the tenacity of Dalit people in the face of oppression, providing a painful glimpse into the complexity of social stratification and injustice. She describes her community's daily struggle for access to water, highlighting how environmental resources played an unjust role in their lives. Kamble and her friends set out to gather water, carrying earthen pitchers on their heads (Kamble 2017). They travel together along the banks of a stream, eventually arriving at a tall rock formation with a cave. A little creek flows through this cave, and people have created a channel to direct the water. They use broken coconut shells to dig into the stream bed, collecting water that seeps through the sand until their pitchers are full. To reach the cave, they must climb steep boulders one at a time, taking turns collecting water and washing their faces in the stream (Kamble 2017).

This depiction highlights the perseverance and resourcefulness required by Dalit women to meet the most basic necessity: clean, potable water. It emphasises the knotted difficulties that their gender identity and caste impose on them. This narrative exposes the unequal allocation of essential resources and the environmental prejudice that marginalised populations face from the perspective of environmental justice. The fact that Dalit women were forced to travel great distances to obtain clean water highlights the systemic inequalities perpetuated by societal structures that prioritise the needs and comforts of affluent groups over downgraded communities.

It is crucial to understand the relation between social injustice and environmental degradation to approach this issue through the prism of environmental justice and discrimination against Dalit women. Due to issues including environmental exploitation and discrimination based on caste, Dalit populations are disproportionately denied access to clean water, a fundamental human right. Dalit women and other oppressed groups suffer from underlying caste-based oppression, which means that any effort to solve environmental injustice must also address this issue. Therefore, any attempt to address environmental injustice must also confront the underlying caste-based oppression faced by Dalit women and other marginalised groups. This requires advocating for equitable resource distribution, challenging discriminatory practices, and amplifying the voices of those most affected by environmental degradation.

Similar to Kamble's literary work, Bama's *Karukku* provides an account of socio-ecological injustice through the lens of a distinct region and caste—Dalit Christians in Kerala. Dalits' rebellion and fight against their long-standing persecution are also described in the narrative. Bama's autobiography exposes the layers of religious hypocrisy and caste injustice, forcing one to confront people's everyday lived realities (Naik 2018). Also, *Karukku* is similar to Kamble's work, in that the writer is not ashamed and considers themselves to be a helpless community. Yet, like Kamble, she believes that we should break free from our victimhood. Bama's book encourages Dalit women to reclaim their identity and place in Indian society through education and entrepreneurship, resisting victimisation by hegemonic powers. The book not only addresses casteism and solidarity among marginalised communities, but also serves as a self-emancipation manifesto for Dalits in India.

In *Karukku*, Bama recounts her experience as a Dalit Christian woman, highlighting how her Dalit identity profoundly shapes her life within the Christian community. She also learns that she must resist discriminatory behaviours both within and outside the Church, a challenge that is particularly difficult for her as a woman in a patriarchal society. However, her increasing consciousness of her Dalit identity marks her entry into the Church and exacerbates her reconsideration and reformation of what it means to be devoted to God, as well as the role of belief and religion in her life. Bama mentions various incidents in her day-to-day experiences as a Dalit Christian woman, but she remains cheerful in asserting "We are never given any honour or dignity or respect. We are stripped of all that. But if we study and make progress, we can throw away these indignities" (Bama 2012).

She narrates a terrific experience of her childhood, when her school teacher unjustly accuses Bama of stealing a coconut and publicly humiliates her, despite the fact that she didn't steal anything (Bama 2012). It was evident to her that the teacher's accusation was based solely on her belonging to the lower *"pariyah"* caste. During her college years, she mentions about the widespread casteist hatred that deeply wounded her impressionable heart: "I had thought that at such a big college, so far away, among so many different students, nobody would bother about such things as caste" (Bama 2012). However, caste distinctions were still a consideration, even in that setting. Furthermore, Bama was horrified to observe that, even when using public transit, the majority of upper-caste women would avoid sitting next to Dalit women (Bama 2012). She questions the morality of these actions, which are rooted in the concept of class-based "varna" superiority.

According to Bama, the village was divided into socio-ecological domains of rivalry and conflict. She observes "The post office, the panchayat board, the milk depot, the big shops, the church, the schools—all stood in [upper-caste] streets" (Bama 2012). This spatial segregation exposes the intersection of ecological apartheid and human rights violations. By relegating Dalits to unsanitary environments while labelling them as "polluted," caste ideology weaponises space to reinforce dehumanisation (Kumar & Mishra 2022).

Bama sheds light on the state of employment and the equitable contributions made by women in managing household expenses while performing chores. She describes how being a *Paraya*[1] is difficult from an early age, as everyone must labour to survive—whether in the fields or working for the Naickers (a dominant community in southern India). When no other work is available, they perform construction labour: hauling stones, gravel, and soil; digging wells; or gathering firewood from hilltops. Since the Naickers own most of the land, the Parayas work in brick kilns, with every household bound as bonded labourers to Naicker families. Bama recounts working alongside her grandmother as a child, noting, "Grandmother is truly a servant." Her grandmother hires labourers for the Naickers, supervises their work, and ensures they receive wages. She wakes early to fetch water and complete household tasks, just as Bama and other women do. They carry millet gruel to the fields in vessels, and when fieldwork is scarce, Bama receives four annas[2] from her mother to give to the gardener.

---

[1] The Paraya/Paraiyars were originally a community of drummers who performed at auspicious events like weddings and funerals. The Paraya were considered to be so low in status that the word "pariah" came to mean "outcast".

[2] *Annas* was a unit of currency in India and 1 anna was equal to 1/16 of one Indian rupee.

Unlike Kamble's village condition, which was on the outskirts of the society, a nasty village, Bama's mentioned hundreds of references about the inherent natural beauty of her village. In the introductory section of her novel, she describes the vivid landscapes: *"a bright red sun, huge and round, would wake up in the east… [making] its way, peering between the trees, glowing, its light spilling and sparkling"* (Bama 2012). This imagery contrasts sharply with the oppressive social realities Dalits face, underlines the disconnect between environmental richness and caste-based exclusion.

Bama's narrative vividly captures Kerala's breathtaking landscape, from its spectacular mountain ranges—adorned with rugged peaks and lush forests—to its broad plains, where foxes roam freely. She paints a dynamic portrait of the monsoon rains, describing how water cascades down slopes to replenish village streams, while lakes and ponds teem with fish, easily caught in strategically placed earthen pots. For Bama, nature is not merely a backdrop but an active, responsive force that evolves alongside human presence, from simple organisms to complex societies. Her perspective challenges traditional nature writing, which often frames nature as passive and external. Instead, Bama's narrative emphasises reciprocity: the villagers plant and nurture crops, respects the rivers that sustain them, and depend on the land's cycles for survival. This symbiotic relationship—where humans, animals, and the environment actively shape one another—reveals her unique vision of nature-self-society engagement.

While praising the splendour of her village, Bama further touches on the caste order, revealing that members of their community collect wood when there is no work available in the field. But this was not the situation for everyone; upper caste people own large fields with boundaries, wells, and pump sets, and they can eat well without having to work hard (Bama 2012). Bama also discusses the difficult mountain jungles where she used to gather firewood, reflecting on the hard realities of low-caste women. She elaborated on the difficulty of her survival in these forests, including the requirement of paying the forester to allow her to get firewood. As she moved through the dense vegetation, she was scratched by twigs and thorns, which frequently resulted in bleeding and tangled hair. Pushing, shoving, and crawling her way through shrubs and briars became regular as she crossed the harsh terrain. Bama's narration exposes how caste hierarchies perpetuate inequalities, particularly in terms of access to natural resources. Caste segregation is imposed mostly on land, in woods, rivers, water bodies, agriculture, villages, cities, occupations, and food. Caste influences how people use and interact with their surroundings, and sustain social stratification and inequality.

2.4 Conclusion

The conclusion of this chapter highlights the significant influence of caste dynamics on individuals' relationships with the environment, exposing the disproportionate denial of resources to Dalit groups, their forced marginalisation into living circumstances, and environmental degradation. The stories of Kamble and Bama eloquently portray the day-to-day issues Dalit women suffer as a result of environmental injustice, from the difficulties of obtaining potable water to the difficulties of surviving in hostile natural environments. Both autobiographies serve as a strong testimony to the Dalit communities' agency and resiliency in the face of hardship. Defying popular myths of victimhood and regaining their position in Indian culture, Kamble and Bama show pride in their identity and ancestry despite having endured centuries of tyranny. However, despite the backbreaking work undertaken by Dalit women, they are often deprived of dignity and respect in society. As shown in her autobiography *The Prison we Broke*, Bama's grandmother serves as a poignant character of the exploitation and resilience of Dalit women, whose labour sustains entire households yet often goes unrecognised and undervalued. Therefore, from this analysis the chapter suggests the need of—implementing policies that prioritise equitable resource allocation for Dalit communities—ensuring access to clean water, safe housing, and sustainable livelihoods—while rigorously enforcing existing laws like the SC/ST Act to hold perpetrators of environmental discrimination accountable. Equally critical is the inclusion of Dalit women in environmental decision-making processes, amplifying their leadership in shaping policies that directly affect their communities. Educational and awareness initiatives must challenge caste-based stereotypes by centring Dalit narratives in ecological discourse, bridging the gap between theoretical justice and lived realities. Beyond policy shifts, transforming environmental justice into tangible action requires collaborative efforts among lawmakers, activists, and scholars to audit caste-disparate impacts of development projects and elevate Dalit resilience as a foundation for sustainable change. Only by confronting caste as a structural determinant of environmental harm can India move towards truly inclusive justice—one that recognises dignity, equity, and ecological security as universal rights.

Acknowledgements We are sincerely thankful to Jyoti Justin for her thoughtful review and comments on this chapter—particularly in the arena of language editing.

References

Bama. 2012. *Karukku.* Translated by Lakshmi Holmström. Oxford University Press.

Behl, Vaishnavi, and Prakash Kashwan. 2022. Intersectional water justice in India. In *Climate Justice in India: Volume 1*, eds. Kashwan, Prakash, D. Raghunandan, and Anju Mishra, 183–200. Routledge.

Bhattacharjee, Somenath. 2023. Deprivation and Social Exclusion of the Dalits: An Anthropological Observation. *Contemporary Voice of Dalit*. https://doi.org/10.1177/2455328x231199839.

Bhimraj, M. 2020. A Dalit Critique of Environmental Justice in India. In *Contemporary Environmental Concerns: Multi-disciplinary Aspects of Environmental Law*, eds. B. Mishra and S. Mishra, 89–120. Bharti Publications.

Bullard, Robert D. 2001. Environmental Justice in the 21st Century: Race Still Matters. *Phylon* 49 (3/4): 151–171. https://doi.org/10.2307/3132626.

Dastidar, Rukmini. 2021. India's Suffering Female Dalit Farmers Have the Most to Lose. *Foreign Policy*. https://foreignpolicy.com/2021/04/13/indias-suffering-female-farmers-have-the-most-to-lose/

Deshpande, Ashwini, and William Darity Jr. 2016. Caste Discrimination in Contemporary India. In *Inequality and Growth: Patterns and Policy: Volume II: Regions and Regularities*, eds. Ravi Kanbur and Joseph Stiglitz, 248–273. Palgrave Macmillan.
Devrim, Eren. 2024. Caste, Gender, and Environmental Injustices in India's Urban Peripheries: Intersectional Inequalities, Decolonial Praxis. In *Environments of Inequality: Crises, Conflicts, Comparisons*, vol. 3, eds. Malini Ranganathan, Prakash Kashwan, and Farhana Sultana, 187–205. Routledge.
Hiwrale, Anup. 2020. Caste: Understanding the Nuances from Ambedkar's Expositions. *Journal of Social Inclusion Studies* 6 (1): 78–96.
Johnson, Taylor N., et al. 2022. Environmental Justice: The Third Pillar of Environmental Communication Research. In *The Routledge Handbook of Environment and Communication*, 2nd eds., ed. Robert Cox and Phaedra Pezzullo, 63–81. Routledge.
Kamble, Baby. 2017. *The Prison We Broke*. Translated by Maya Pandit. Orient Blackswan Private Limited.
Kumar, Vijay, and Binod Mishra. 2022. Environmental Casteism and the Democratisation of Natural Resources: Reimagining Dalit Testimonies. *South Asia: Journal of South Asian Studies* 45 (3): 577–595.
Learn About Environmental Justice | US EPA 2024. US EPA. https://www.epa.gov/environmentaljustice/learn-about-environmental-justice
Naik, Purnachandra. 2016. Baby Kamble to Bama: Dalit Women Write Differently. *Economic and Political Weekly* 51(42), 16–19. http://www.jstor.org/stable/44165812
Naik, S. 2018. Mahasweta Devi and Bama Voices Against Patriarchy and Society: A Comparative Study. PhD thesis, Kuvempu University.
Naudet, Jules. 2008. 'Paying Back to Society': Upward Social Mobility Among Dalits. *Contributions to Indian Sociology* 42 (3): 413–441.
Oza, Preeti. 2020. Religious and Community Discrimination: Literature as Protest in Dalit Writings in India and Black American Literature in the West. In *Kala Pani Crossings, Gender and Diaspora Indian Perspectives*, Misrahi-Barak, Judith et al. Routledge.
Pasic, Damir. 2020. UN and Dalit Women. *International Dalit Solidarity Network*. https://idsn.org/key-issues/dalit-women/
Prasad, Indulata. 2022. Towards Dalit Ecologies. *Environment and Society* 13 (1): 98–120.
Ranganathan, Malini. 2021. Caste, Racialization, and the Making of Environmental Unfreedoms in Urban India. *Ethnic and Racial Studies* 45 (2): 257–277. https://doi.org/10.1080/01419870.2021.1933121.
Ranganathan, M. 2022. Caste, Racialization, and the Making of Environmental Unfreedoms in Urban India. *Ethnic and Racial Studies* 45 (2): 257–277.
Rao, Nitya. 2014. Caste, Kinship, and Life Course: Rethinking Women's Work and Agency in Rural South India. *Feminist Economics* 20 (3): 78–102.
Rege, Sharmila. 2014. *Writing Caste/Writing Gender: Narrating Dalit Women's Testimonies*. Zubaan.
Sharma, Mukul. 2012. Dalits and Indian Environmental Politics. *Economic and Political Weekly* 47 (2): 46–52.
Sharma, Mukul. 2019. 'My World Is a Different World': Caste and Dalit Eco-literary Traditions. *South Asia: Journal of South Asian Studies* 42 (6): 1013–1030.
Sharma, Mukul. 2022a. Caste, Environment Justice, and Intersectionality of Dalit-Black Ecologies. *Environment and Society* 13 (1): 78–97.
Sharma, Bhushan. 2022b. Dalit Women: Narratives of Vulnerability, Violence, and a Culture of Impunity. *Journal of International Women's Studies* 23 (6): 7–19.
Simon, Surya. 2021. Transformation from Aesthetics to Activism: An Analysis of Select Dalit Women Personal Narratives. *Contemporary Voice of Dalit* 15 (1): 109–122. https://doi.org/10.1177/2455328x211038986.
Srinivas, M.N. 1957. Caste in Modern India. *The Journal of Asian Studies* 529–548.

Srinivasan, Prakash, Prakash Kashwan and Ashlesha Khadse. 2022. Excerpt: How a Women's Collective Fought the Climate Injustice Wracking India. *The Wire*. https://science.thewire.in/politics/rights/climate-injustice-india-tn-women-collective

Tamil Nadu Untouchability Eradication Front [TNUEF]. 2019. Water and Caste Exclusion: Survey Findings from Urban Tamil Nadu.

The Dalit: Born into a Life of Discrimination and Stigma. 2021. United Nations Human Rights Office of High Commissioner. https://www.ohchr.org/en/stories/2021/04/dalit-born-life-discrimination-and-stigma

Vishwakarma, Aarti. 2022. Depiction of Plight and Subjugation of Dalit Women in Baby Kamble's *Prisons We Broke*. *The Creative Launcher*, June. https://www.semanticscholar.org/paper/Depiction-of-Plight-and-Subjugation-of-Dalit-Women-Vishwakarma/8280d97742b116df7bb9f31ae99b8f61a94bf79f

Monika is a PhD student at the School of Humanities and Social Sciences and the Indian Institute of Technology Indore, India. She is a visiting research scholar at the Department of Sociology at Boston University, USA. Her research interests include Democracy, Caste Politics, and Nationalism in India.

Mohd Amin Khan is a PhD student at the School of Humanities and Social Sciences and the Indian Institute of Technology Indore, India. He is a visiting research scholar at IIRS ISRO. His research area includes wildfire dynamics, focusing on its impact on tribal communities in India, and religion among indigenous people.

Pritee Sharma is a Professor at the School of Humanities and Social Sciences at the Indian Institute of Technology Indore, India. She is also the Lab Head of the Sustainability Studies Lab at IIT Indore India.

Chapter 3
Voices of the Everyday: Dalit Aesthetics and Ecology in Select Short Fiction from Malayalam

Dhanesh Mankulam

Abstract This chapter explores the intersections of Dalit aesthetics and Environmental Casteism through a close reading of two Malayalam short stories in translation—T. K. C. Vaduthala's 'Sweet-offering at Chankranthy' and M. K. Madhukumar's 'Paalakkunnan's Journey', both featured in *The Oxford India Anthology of Malayalam Dalit Writing* (2012). Based on these stories, the chapter argues that Dalit aesthetics, rooted in the lived realities of caste oppression, offers a distinct narrative mode that challenges dominant Savarna frameworks of representation of nature—particularly those that aestheticise nature as an object of leisure, stylisation, or romantic abstraction. In contrast, Dalit representations of nature emerge from positions of marginality, labour, and survival histories. Nature is not separate from life but enmeshed in the daily struggles of land, body, faith, and community. This embodied ecological consciousness questions the artificial binaries between nature and culture, aesthetics and politics, and subject and environment. The stories examined in this chapter foreground how caste identities shape one's relation to the environment, revealing how ecological experience is stratified and mediated by structures of power and exclusion. By engaging with the everyday as a critical site of environmental and aesthetic articulation, this chapter argues that Dalit literature offers a powerful counter-discourse to dominant ecological imaginaries. It asserts the importance of reading environmental questions through the lens of caste and everyday life, thereby contributing to the growing discourse on Environmental Casteism and the politics of representation in Indian literature.

Keywords Environment · Dalit aesthetics · Inaesthetics · Alternative · Dalit literature · Everyday

---

D. Mankulam (✉)
Faculty of Arts and Humanities, CHRIST University, Bangalore, India
e-mail: dhanesh.mankulam@gmail.com

S. Kochar and R. Singh (eds.), *Pollution and Poverty*,
https://doi.org/10.1007/978-981-95-6876-5_3

3.1 Introduction: Concept of Aesthetics

The traditional schools of Aesthetics considered aesthetics to be the study of beauty and taste. It concerns with the appreciation of beauty in general too. As a discipline, Aesthetics involves the careful study of sensory experiences, such as how we perceive and respond to visual, auditory, and other sensory stimuli, and explores what makes something aesthetically pleasing or meaningful. Aesthetics can also refer to the set of principles or standards used to evaluate and appreciate art, design, or other forms of expression. As Kant (1790) discussed in the *Critique of Judgement* (1790), the attempt was to analyse certain states of mind involved in the very faculty of appreciation of beauty. Edmund Burke (1757), on the other hand, aimed at the language and concepts involved in the aesthetic criticism. Immanuel Kant, David Hume, and others pioneered the discourse in earlier times. In recent times, a variety of discursive vectors have been developing. These vectors have developed in close correspondence with existing discourses on culture, Dalit and minority experiences, gender and subjectivity, and socio-political dimensions. These dimensions include Late Capitalism, New Leftism, Everyday Life, fashion, etc. However, aesthetics has been considered a matter of appreciating beauty and taste in art and elsewhere. This chapter intends to extend this definition tied to art appreciation to a broader meaning.

The discourse surrounding culture, life, and ideology in contemporary discussions needs a comprehensive and ongoing exploration of a fundamental aspect crucial to the field of politico-cultural studies: the human faculty of appreciation. This faculty can be counted as one of the many qualitative metrics of one's social life. Broadly put, aesthetics deals with this human faculty of appreciation, traditionally understood with reference to art and such creative production, but by extension, in everyday decisions as well. Stephen Kaplan, a prominent American psychologist and scholar widely recognised for his work in environmental psychology, studied the environmental preference of humans through an evolutionary lens, arguing that humans have developed an innate tendency to favour certain landscapes due to their survival benefits: 'Scenes of the outdoor physical environment vary substantially in the extent to which they are preferred' (Kaplan 1987). Kaplan suggests his 'framework for environmental preference', where unconscious cognition may occur prior to certain affective judgments. The same point can be found relevant to the sociology of aesthetics as well. That is why the notion of appreciation goes beyond the objects of art and literature, gaining much broader socio-political implications within one's everyday experiences. From the perspective of Cultural Studies, cultural factors are inseparably tied to the biological aspects of human beings. An act of appreciation pertains to a judgement of value, a determination of the normal, an evaluation of merit, and a general conviction about the desirable in a social ensemble. The questions of equality, equity, accessibility, and identity revolve around these aspects of the act of appreciation or aesthetics. In other words, practices of equality, equity, accessibility and identity involve factors of preferences as well.

A deeper examination reveals that the act of appreciating and evaluating the world is intrinsic to the existence of our species as a socio-political and cultural collective. In that sense, 'aesthetics' must also be understood within the 'sociology of aesthetics'. The sociology of aesthetics refers to how the varied factors involved in an aesthetic evaluation are also reflective of the different rules, privileges, norms, systemic violence, and hierarchies of that society. Scholars have identified the potential of affective aesthetics to cultivate 'habits of mind' (Bresler 2006). This 'habit', which is an institutionalised disposition towards certain aspects of social life in terms of right, legit, superior and normal, also implies that every systemic injustice is also a habituated injustice. The caste system is one such example. An aesthetic engagement becomes a political engagement when we try to look deeper into this sociology of aesthetics. It answers questions pertaining to why and how something is valued in a certain way, who determines and reiterates a particular value of a structure/person/community/act, and how something like a body feature is kept as the desirable standard, which in turn reciprocally defines and/or defiles every difference. Stephen Kaplan views that 'aesthetics plays a significant role in human behaviour and experience, with rapid, unconscious cognition potentially preceding certain affective judgments…' (Kaplan 1987).

The French philosopher Jacques Rancière introduced an expression, 'the distribution of the sensible'. The term 'distribution of the sensible' refers to the system of self-evident facts of sense perception. According to Rancière, this system not only reveals the existence of something common to all but also establishes the boundaries that define specific roles and positions within it. As Rancière explains, 'a distribution of the sensible therefore establishes at one and the same time something common that is shared and exclusive parts' (Rancière 2004, p. 12). This idea highlights how the perception of the world is shaped by both what is universally shared and what is divided, offering a framework for understanding how social and cultural boundaries are formed.

These 'self-evident facts' function like touchstones in all socio-political contexts. It is against these self-evident facts that one has to try every statement of judgement. In everyday communication, we refer to this as 'common sense'. Various socio-political aspects such as objects, spaces, times, and forms of activity are distributed along this common sense. According to Rancière, these facts set the various 'forms of visibility' (2004, p. 13). Forms of visibility are how one normatively sees the world and, in turn, is normatively seen by others. Hence, forms of visibility are also forms of perceptions: it institutes specific ways for oneself and others to appear to everybody's perception. When these forms of visibility function in favour of some groups of society—when their needs, appearance, features, and preferences become normalised within these forms of visibility, when some feel at home while others feel alienated within these forms—that exposes the real politics of perception.

In a broader sense of the term aesthetics, these politics of perception underlie our basic acts of appreciation. Aesthetics is not any neutral or objective evaluation of something. It adds to the diverse discourses surrounding different domains, meanings, and items that are appreciated. There is a methodological ground to do it. Every human collective is founded on some commonly applicable standards of appreciation—the social structures of aesthetics. The process of appreciation serves as both an overt experience and a covert force. It operates not solely as the cohesive element driving what can be referred to as the experiential existence but also as a foundation for all the semantic, hierarchical, and functional offshoots of the socio-politico-systemic ensemble, including aspects such as caste, class, and gender. The logic and values of the prevailing caste system in India—where the Brahmin groups come on top of the hierarchy, and the Dalit groups come at the bottom—have always been maintained and mediated based on a particular aesthetics which functions at both de facto and de jure dimensions to present the aestheticism of the so-called upper castes as the normal and the model. The distance from the upper caste along the caste hierarchy is then defined not only in terms of existential compromises but also that of social aspirations.

Adrian Currie, a scholar known for his work on areas of perception, cognition, and the role of aesthetics in knowledge formation, believed that aesthetic engagement is not just a passive experience but also a constitutive one. According to him, it serves as the locus of justification for all kinds of epistemic activities (2023). An aesthetic evaluation sets in motion a set of values, identities, and practices which constitute the social experience and reality of an individual or community. The idea of an alternative aesthetics emerges from a problematisation of the existing aesthetics, which constitutes and maintains the existing social system in a particular way with specific values, identities, and identity practices. Dalit aesthetics initiate a revaluation and reconstitution of meanings and values that run the existing caste systems and Brahminical logic. It is a conscious attempt to relook at and reimagine all categories of knowledge in terms of creative expressions, historiography, epistemology, ethics, tradition, and beauty. Adrian Currie recognises the relationship between the aesthetic and the epistemic, and the term he uses is 'attunement': 'scientific training (and other knowledge-directed activities), I argue, involves "attunement": the co-option of aesthetic judgements towards epistemic ends' (Currie 2023).

3.2 Concept of Dalit Aesthetics

The idea that underlies a proposition of Dalit aesthetics is that there is what can be called a 'regime of aesthetics'. Michel Foucault, a prominent French philosopher known for his theories on power, knowledge, disciplinary societies, and social institutions, talked about 'regimes of truth.' He explained that these regimes are 'that which constrains individuals to a certain number of truth acts' (Foucault 2014, p. 93). By this, Foucault means that a regime of truth 'defines, determines the form of these acts, and establishes their conditions of effectuation and specific effects' (2014, p. 93). Furthermore, it also 'determines the obligations of individuals with regard to procedures of manifestation of truth' (2014, p. 93). In other words, these regimes of truth determine the ways in which individuals are normatively compelled to express and interact with truth. This framework parallels the idea of a 'regime of aesthetics', where the norms of aesthetic perception are similarly structured and controlled. Thinking along the same lines, a regime of aesthetics implies those 'formulae' for aesthetic appreciation which institute a specific pattern of judgement in a social group. By limiting the individuals to its set of formulae, an aesthetic regime determines and regulates the individual acts of aesthetic appreciation. Establishing the conditions of the occurrence and recurrence of certain types of aesthetic appreciation establishes an obligation on the members of the social ensemble to perform their identities and expressions based on the given conditions of appreciation. It also limits, segregates, and appropriates the language and all other aspects of articulation in its own favour.

Jacques Rancière (2004) distinguishes three artistic regimes: the ethical, the representational, and the aesthetic. Under the ethical regime, he places Plato's moral philosophy of art. Under the representational regime, he places Aristotle and others. Under the aesthetic regime, he groups all the different tributaries of Modernism. Rancière talks about an aesthetic regime of politics in which 'politics revolves around what is seen and what can be said about it, around who has the ability to see and the talent to speak, around the properties of spaces and the possibilities of time' (2004, p. 13). However, this chapter would state that the point in consideration is not just the aesthetic regime of politics; it is also about the socio-political engagement of a general regime of aesthetics itself. When this chapter uses the expression 'regime of aesthetics', it draws from both Ranciere's notion of an aesthetic regime of politics and Foucault's notion of a 'regime of truth'. It begins with the following questions: How does everything come into sight of the everyday perusal of society, why specifically so, and what are the other possibilities of coming into sight?

The proposition of Dalit aesthetics directly addresses the existing regimes of aesthetics that validate the Brahminical ways of life while maintaining various stereotypical representations of Dalit life, experiences, and realities on a logic of alterity. Dalit aesthetics begins where a 'Brahminical regime of aesthetics' is identified and problematised. It critiques the existing models in literature, social theory, philosophy, cultural representations, nationalism, popular media stories, archival practices, historiography, and epistemology as being the production sites for a Brahminical casteist regime of aesthetics. Dalit aesthetics then attempts not only to disengage with this Brahminical regime of aesthetics but also to establish sites for alternative and critical social productions. This movement was initiated in literature and cinema. Critically responding to this Brahminical regime of aesthetics, Manju Edachira, a scholar of Dalit studies, notes, 'Dalit literature largely remained as an alternative, if not an opposition, to mainstream literature' (2024). While most evident in art, literature, and cinema, this Dalit alternative also functions as a critical positionality against the mainstream perspectives in society. Critiques that emerge from literature and cinema are debunking the violence and politics of the systemically embedded casteist perceptions.

A politics of oppositionality is at the centre of Dalit aesthetics. It is a 'look-back'. Alok Mukherjee, a prominent scholar and translator of Dalit literature and critical theory, while introducing his translation of Sharankumar Limbale's 'Towards an Aesthetic of Dalit Literature' (2004), says, 'Dalit writers have insisted that their writing has a particular purpose and audience, … and their works should not be assessed by "universal" criteria, which, in India, carry the markers of caste and class' (Mukherjee 2004, p. viii). This 'universal' that Mukherjee mentions is the idea of the 'regime of aesthetics' in consideration here. Noting how these 'universal criteria' devalue Dalit writings as lacking literary merit, Mukherjee draws the focus to Limbale's observation that 'Dalit reality is a liminal reality in time and space' (Mukherjee 2004, p. ix). One purpose of Dalit aesthetics would then be to reorient the whole domain of aesthetic appreciation—not just in art and literature, but in social recognition also—from the centrality of the existing regime of aesthetics to a perspective from Dalit social experience. This move is not about a speech from the margin; it is about realising a centre in what is otherwise deemed marginal by the mainstream. K. Ayyappa Panicker, a literary critic from Malayalam, commented on this distinction, stating, 'While classical arts manifest the meaning of life through extensive and intensive study and decorative expressions, folk cultures imbibe the pleasures and pains of base lives with the power of nature' (Panicker 2022, p. 58).

3.3 Dalit Aesthetics: Characteristics

The factors characterising Dalit aesthetics must be understood with this premise. One of the key characteristics of Dalit aesthetics in literature and cinema is its affective and expressive quality. Because of the historical and institutionalised discrimination based on caste hierarchy, Dalit experiences and realities were often overlooked by the creative expressions of the 'regime of aesthetics'. The Brahminical aesthetics always focused on metrical composition and exaggerated styles at the cost of real-life experiences and everyday realities because struggle and survival were far from their primary social reality. 'The basis of Indian aesthetics was "stylised acting". That is always an exaggeration as well. It used sign movements instead of natural movements; it understood reality in symbolism… It even turned logical propriety into just embellishment' (my trans., Pambirikkunnu 2007, p. 27). The logic mentioned here implies a constructed thought pattern, such as in terms of language, symmetry, etc., almost like an abstract Kantian pure reason. This 'tyranny of logic and artificiality' characterises the Brahminical regime of aesthetics. As Manju Edachira notes, 'the necessity of rational epistemological questions undermined the affective expressive aesthetics of literary works' (2024). This assertion rightly draws attention to how dominant aesthetic frameworks often privilege reason over emotion, but it risks overlooking how affect and expression can themselves operate within, or even subvert, rational forms.

As against this 'tyranny of logic and artificiality', the everyday toil, non-standardised expressions, and functional and communicative language had become unfit to receive any special aesthetic gaze. An extension of the same approach made labour, workers, soil, sweat, ordinary expressions and actual lived realities—all connected to a working-class life—undeserving of attention. Dalit aesthetics proposes to bring to the front the latter: the everyday toil, non-standardised expressions, and functional and communicative language along with labour, workers, soil, sweat, ordinary expressions, and actual lived realities. It is more affective and expressive in nature, less standardised, less reductively rationalised, and less artificially embellished: an affective expression of an actually lived life. Sharankumar Limbale observes, 'Dalit writers believe that Dalit literature is a movement. They see their literature as a vehicle for their pain, sorrow, questions, and problems' (2004, p. 105). Dalit aesthetics then identifies and responds to that gap in the mainstream literary aesthetics—the Savarna aesthetics in Limbale's words—which prescribes objective standards of literary criteria to guide any act of literary appreciation. The same applies to the general category of aesthetic appreciation as well. The Savarna aesthetics in literary criticism that Limbale (2004) recognises is the literary counterpart of the 'regime of aesthetics' in the broader social scenario. If 'Dalit literature is life-affirming literature' (Limbale 2004, p. 105). Dalit aesthetics proposes an affirmation of Dalit life-experience across social interactions and exchanges.

Another aspect characterising Dalit aesthetics is 'inaesthetics'. Alain Badiou (2004), a prominent French philosopher known for his work on ontology, politics, and philosophy, describes 'inaesthetics' as a relation of philosophy to art that asserts art itself is a producer of truths, without being reduced to an object for philosophical analysis. According to Badiou, 'against aesthetic speculation, inaesthetics describes the strictly intraphilosophical effects produced by the independent existence of some works of art' (2004, p. xii). Badiou's proposition is about a different way of understanding the relationship between philosophy and art. Instead of treating art as something to be analysed and dissected by philosophy as in the traditional schools of aesthetics, 'inaesthetics' suggests that art itself is a generator of truths. This is one of the convictions of Dalit aesthetics as well. Pradeepan Pambirikkunnu, a scholar of Dalit aesthetics, literature, and cultural criticism from Malayalam, observed that the traditional aestheticism in India was merely constructing romanticised generalisations about reality instead of understanding its particularities, and such romanticisation ignored those who lived in the mundane particularities of life (2007, p. 28). Dalit aesthetics rejects this practice of romanticising realities. It is in this rejection that it fosters 'life narratives' where lived reality of life becomes the content, form, and truth of expressions. Dalit aesthetics refuses to validate the presentation of lived realities based on given normative standards. As Parthasarathi Muthukkaruppan, a scholar of Tamil Dalit literature, rightly observes, the questions raised by Dalit aesthetics are 'historical in nature and warrant historical explanations rather than abstract normative ones' (2017, p. 64). The highlight is the repositioning of oneself from the role of a consumer of aesthetic values to that of a producer. For instance, Muthukkaruppan notes that in Bhama's *Karukku*, 'while the community still occupies the side of physical labour within the ideological grids of the division of physical and mental labour, through remembering the past, the author re-members herself in the community' (2017, p. 67).

Elements pertaining to physical life become central to Dalit creative expressions within this logic of inaesthetics. While discussing the aestheticism of the classic Malayalam novel *Indulekha* (1889), Pradeepan Pambirikkunnu (2007) notes that the traditional Malayalam literary aestheticism continued to present beauty in bizarre imaginations. It ascribed letters with images of wheels and snakes and even gambled with the last lines in the name of 'open questions'. It continued to celebrate the realities of life as a stylised linguistic acrobat of pure aesthetics. The primary preoccupation for both the readers and the author of the novel *Indulekha* was a moral anxiety over where one's dress ends and the body begins. It was indulging in the visual beauty of the body based on certain mainstream norms of aesthetic judgements (Pambirikkunnu 2007, p. 28). Dalit aesthetics, especially in the literary domain, focuses on physical labour and material social relationships in the 'language of common people without sanitisation' (Muthukkaruppan 2017, p. 67). Such lives are not beautifiable with artificial figures of speech (Pambirikkunnu 2007, p. 28). Instead, 'Dalit life narrative… appears to present itself as a life narrative of a community or a collective life' (Muthukkaruppan 2017, p. 68). What Dalit aesthetics offers is 'a new thinking and a new point of view' (Dangle 1992, xii).

Material social '"situatedness" characterises Dalit aesthetics. The experiences articulated in Dalit literature have not yet been expressed in any other literature. They are the experiences of a particular community …its character is collective rather than individual'. (Limbale 2004, p. 31). Dalit aesthetics approaches the world and its normative discourses from this situatedness. Body, for example, in Dalit literature, is a body that is put to the test of social and physical labour. As L. S. Rokade, a Marathi Dalit poet and critic, writes, a '… body covered with generations of dire poverty' (Rokade 1992, p. 1). It is presented as a site of social, historical, and political clashes, a body that is dialectically and materially positioned against the mainstream. Its posture is that of physical labour, its setting is that of land, soil, and community, and its engagement is that of emotion, affect, and coexistence. It doesn't strive to appear like an objective philosophical body linked to stylised language, abstract reasoning, and universal anxieties. The mainstream Savarna aesthetics is irrelevant to describing such a body because, to use terms from Limbale, for the Savarna aesthetics, this body is linked to 'untruth', 'unholy', and 'unbeauty' (Limbale 2004, pp. 20–21). Hence, Dalit literature need not subscribe to the idea of 'timeless literature', which characterises traditional literary criticism. Rather, it must consciously bring time and history into its literature and expressions.

3.4 Dalit Aesthetics in Dalit Literature

Dalit literature has been instrumental in the development and popularisation of an aesthetic alternative that can be called Dalit aesthetics. The two stories in consideration for this study—'Sweet-offering at Chankranthy' by T K C Vaduthala (2012) and 'Paalakkunnan's Journey' by M K Madhukumar (2012)—demonstrate how a Dalit subject position plays the role of a narrator. These stories are taken from *The Oxford India Anthology of Malayalam Dalit Writing* (2012), edited by M. Dasan, V. Pratibha, C.S. Chandrika, and Pradeepan Pampirikunnu. Both stories deal with the everyday toils of the Dalit working-class community through non-standardised expressions and a language that is really used in day-to-day communications. Quite uniquely, the voice of the narrative, rather than being that of a particular individual, extends to represent the life of a collective. 'Dalit orality is pronouncedly concerned with an experience that is not individualistic but is always collective' (Dasan et al. 2012, p. xxvi).

The story 'Sweet-offering at Chankranthy' begins with a description of the setting where the agriculture song rises up from the rice field. The *njaatu* song, as it is called in Malayalam, sprung not from any leisure or luxury but alongside the everyday labour in the rice fields and in a language of rhythm rather than metrical precision. The singers are all standing knee-deep in the mud. Their labour is the 'everyday ritual' of their collective existence. The subject of their rhythmic song is not any universal subject. The song is about themselves, their life events, and their emotions. It is a song about their collective survival. Through their songs, they not only narrativise their otherwise unrecorded lives but also share the emotions, pains, and passions that bind them together as a social collective. This close link with lived life is the beauty of these songs. Such thrust on everydayness can be seen in other communications made by the characters, especially between the characters Vellon and Kochukarumbi (Vaduthala 2012, pp. 58–59).

The story 'Paalakkunnan's Journey' begins with a description of the place: Paalakkunnu, the village perched on the hill where the ancient *Paala* tree (Alstonia Scholaris) once stood tall, was now almost dead. The *Paala* tree holds profound spiritual meaning, often symbolising deep traditions of coexisting with nature, as seen in many Indian stories and folktales. It also represents the continuity of life across generations, embodying a living connection to the past. The great tree, which had once soared into the sky and cast its expansive green canopy across the land, now lay in total ruins. Paalakkunnu, bound to the roots of this dying tree, 'clung to the roots that had grown across the cycle of innumerable births and deaths' (Madhukumar 2012, p. 81). The once vibrant source of life, now a shadow of its former self, still held on, as the tree 'held its breath, smothering the pulses of generation' (Madhukumar 2012, p. 81).

The tinge of anthropomorphism gives an impression of the tree, the hill, the village, and the people sharing one context of life. This ecology is central to the Dalit imagination of the world as the communities are socio-historically linked to nature by labour. Moreover, the tree and the hill are witnesses to the existential pulses of generations. For the people working in the fields, knee-deep in mud, these ecological factors are not symbols of romanticism but integral parts of their lived reality. A Dalit person cannot look at the ecological factors as distant and stylised as in the classical Savarna literature. As the story suggests, the history of the Palakkunnu Hill and the great tree on it is also the history of the people who live there. That history is not a written one but a lived and shared one. Humans and nature are participants in that history. It exists in the collective disposition of their life. 'Collective' is the quintessential code of their socio-political and ecological reality.

The argument here is not that the stories somehow lack a philosophical underpinning. On the contrary, as the editor comments in the General Introduction section of the collection *The Oxford India Anthology of Malayalam Dalit Writing* (2012), there is a certain Dalit cosmogony which is not absolute: 'rather, this works as an unconscious knowledge, as an unconscious background information that structures the narrative of faith' (Dasan et al. 2012, xxviii). On one of the occasions in the story 'Sweet-offering at Chankranthy', Vellon plays a prank on Kochukarumbi by taking away the food she offered to her dead husband on the auspicious day of 'Chankranthy'. She could not offer food to the dead regularly because of her poverty. Hence, she thinks the dead ate all the food as that soul must have been starving. She is heartbroken, and she cries, '*Ente Taive!* My God! *Ittaram porinchaanu* ... so famished ... *aen arinjchillalla,* I didn't know! Would've borrowed and got something more ... *kado velemedicchu enthenkilum konduvanthu tanthaanella!*' (Vaduthala 2012, p. 60). The whole intention here is not an alienation of nature or the supernatural as 'other'. Rather, it reflects the deep connection between the living and the dead, where the act of offering food transcends materiality and becomes a bond of care and empathy. Kochukarumbi's reaction emphasises the human tendency to assume the suffering of others, even beyond death, and the guilt she feels for not doing enough, even when her circumstances prevent her from doing so. In Dalit aesthetics, the connection between the material and the immaterial is not viewed from a rational detachment. Instead, it is felt viscerally—through embodied experiences of struggle, care, and connection. This visceral experience forms the essence of Dalit aesthetics, as reflected in this story.

The translator also used the source language alongside English to share the sentiment. This scene is an intersection of the material reality of poverty and a unique cosmogony. The philosophy of life here is not a mere rational absolutism expressed in a specialised stylised diction. This is integrated with the lived realities' everydayness that characterises Dalit literature and, in turn, Dalit aesthetics. The 'tyranny of logic and artificiality' is overcome here. The philosophy itself is functional and communicative. This 'inaesthetic' of Dalit literature refuses to separate the rational from the emotional. As a result, art and literature are no longer merely objects of philosophical contemplation but are redeemed in their own right. The historical sense in Dalit literature is not confined to rational empiricism. A community historically marginalised by dominant historiography cannot rely on the methods of positivist histories. Instead, the history here is rooted in the lived experiences of the place, the plants, and the people. It is a history that intertwines the environment, culture, and the struggles of those who have been left outside the dominant narratives.

Elements pertaining to physical and natural life are central to Dalit creative expressions and 'Dalit inaesthetics' as an alternative to the mainstream aesthetics. As noted earlier, nature in Dalit literature is not the nature in William Wordsworth's poems. Dalit literature does not alienate nature, nor does it artificially stylise it. Nature is always found entangled with lived life. Dalit literature demonstrates that through the affective and expressive qualities of its narratives. The song at the beginning of the story itself is one example of this expressive and affective quantity. Rather than conforming to universal themes, it prioritises the visceral realities of existence and the affective bonds binding individuals together. The toil, the mud, the vastness of the field, and the rhythm of the song are all part of the same aesthetics of a Dalit worldview. By foregrounding the physical and emotional dimensions of marginalised existence, Dalit writers foster a deep sense of collective identity and solidarity.

The sense of collective identity and solidarity develops from a recognition that the existing narratives—of history, legends, scriptures, ethics, and aesthetics—are all created from the Savarna point of view. In the story 'Paalakkunnan's Journey', a Guru (teacher) begins to narrate a story of their ancestors to the younger generations. The story was that of Maveli, a legendary 'asura' King who was known for his exemplary adherence to justice. The legends narrate that Lord Vishnu, in the form of a Brahmin boy, pushed Maveli to the netherworld to satisfy the envious Devas. The big festival of Onam in Kerala is based on this story. As the Guru mentioned Maveli, the children responded, 'But Vaamana sent him to the netherworld' (Madhukumar 2012, p. 83). The Guru replied in rage, 'That is what Vaamana and his cohorts will always say. ... No Vaamana can ever send Truth to the netherworld' (Madhukumar 2012, p. 83). The Guru represents the alternative perspective of Dalit aesthetics, historiography, and narratology, where the Savarna narratives are held accountable for their historical erasures, injustices, and perpetuation of caste-based hegemony. This narrative is not a rational claim like a modern historical account; it is the affective expression of the community's historical resentment. 'Deconstruction of these kinds of myths is essential as a counter to the portrayal of the indigenous population as demonic' (Kumari and Kapoor 2021, p. 7).

This emphasis on affective expression can be felt rather than seen in many of the interactions among the characters. 'It was the *Chankranthy* day in the month of karkatakam. There is no day more sacred than this for the Pulayas' (Vaduthala 2012, p. 59). This day is regarded as highly auspicious by many communities in Kerala, with people performing rituals in honour of the dead and their ancestors. 'It is on this day that all those who had departed were offered *taakam*. ...Kochukarumbi too did not forget this. She had kept apart a little from the wages she got each day' (Vaduthala 2012, p. 59). Faith is not a sole abstraction here. The narrative presents the experience of faith within a material condition. The socio-economic-material reality of life is inseparable from the affective and emotional reality. Reality is conceived as physical, lived, and affective. Pramod K Nayar, a scholar and critic known for his work on postcolonial studies and cultural theory, uses the term 'traumatic materialism' to refer to the different sensibility that Dalit writings demand: the aesthetic of survival emerging a historically traumatic socio-economic condition (Nayar 2015, p. 1).

Dalit aesthetic finds itself in the material, social 'situatedness' of the Dalit people' (Dasan et al. 2012, p. xxvii). Dasan elaborates that this social positioning also links them to the ecological context through their labour. Hence, the term 'Dalit' not only signifies the unique socio-political identity of a Dalit individual, which is largely excluded from mainstream classical art and literature, but also encapsulates the entire social reality of Dalit life, including labour, relationships, emotions, and faith. As Dasan describes, the songs of Dalit life are 'earthy' (Dasan et al. 2012, p. xxvii). It embraces those realities of life—of labour, mud, pain and faith—that the 'savanna aesthetics' recognises as the 'other' unfit for literary appreciation. 'The life experience of the spectator of a *Kathakali* or *Kootiyattam* is not conditioned by the angst of survival' (Dasan et al. 2012, p. xxv).

This lived connection contrasts with Savarna aesthetics, which often romanticises nature in the Wordsworthian style. On the contrary, Dalit aesthetics looks at nature not as a beautiful ‘other’ but more as an extended self. As the renowned Indian scholar GN Devy notes, ‘[the subaltern communities] accept a worldview in which nature, man and God are intimately linked’ (Devy 2002, p. x). Nature is not an object of contemplation but an intimate presence deeply tied to existential realities. The classical Sanskrit aesthetic tradition also, particularly in the theories of ‘rasa’ and ‘dhvani’, often presents nature as an aestheticised landscape meant for elite enjoyment. Bharata’s *Natya Shastra* and Abhinavagupta’s commentaries prioritise a metaphysical appreciation of beauty, where the natural world is celebrated in poetry and drama but severed from the material conditions of labouring bodies. Dalit aesthetics challenges this by embedding nature within the politics of survival and daily life rather than framing it as a passive backdrop for artistic contemplation. As seen in texts like Kalidasa’s *Meghadūta* or Jayadeva’s *Gita Govinda*, nature is rendered in sensual and devotional imagery to make it extraordinary and sometimes surreal. Such rendering demands an aesthetic sensibility and ideological lens that may be termed Brahmanical and elitist in nature. This type of appreciation frequently abstracts nature from its labouring subjects, reinforcing an aesthetic of leisure that is accessible only to the upper castes. In this way, ecology becomes linked to leisure, imagination and surreal creativity.

Dalit ecological discourse disrupts this elitist tradition by asserting a counter-aesthetic of labour, where mud, sweat, and survival replace leisurely detachment and refinement as primary modes of engagement. As Indulata Prasad, a scholar focused on environmental justice and Dalit studies, argues, ‘Dalits are disproportionately impacted by natural disasters and climate change because of their high dependence on natural resources and manual labour, including agriculture’ (2022, p. 1). The dominant environmental discourse, often coloured by the aesthetics of leisure and the mainstream casteist codes, tends to overlook these intricate ways in which caste identity and ecological symbols intersect. She draws attention to how Dalit writer Om Prakash Valmiki (1950–2013) poignantly described enduring forms of Dalit dispossession: ‘Using ecological terms such as mud, pond, food grains, fields, harvest, water, and well, Valmiki noted that all material resources belong to the landlord, then questioned where Dalits might belong in society’ (Prasad 2022, p. 106).

Building on Valmiki’s exploration of Dalit dispossession, it becomes evident that Dalit literature consistently portrays a deep, almost inseparable connection between Dalit identity and the ecological realities they inhabit. The selected stories depict an alternative framework by highlighting the intricate entanglement of nature (in terms of language, symbols and experience) with the unique Dalit identities. The word ‘chankranthy’ used in the story is an alteration of the more standardised ‘Sankranthy’ that is used in the formal print form of Malayalam. The former culturally represents the language of the unlettered, while the latter reflects the language of the learned. The Dalit alternative aesthetic of language begins here, in this very act of language use. Furthermore, the subjects—a simple rice offering made during Sankranthy and the tale of ‘paalakkunnu’, a hill with a tree atop—seem like topics that would be dismissed as too trivial for ‘high literature’ by the traditional Brahmanical aesthetic sensibilities.

3.5 Conclusion

The classical art and literature in India resemble the European neoclassicism and later Romanticism. It defines aesthetics and literary or artistic quality in terms of artificially stylised presentations. Even the presentation of nature is often romanticised to remove every aspect of real lived experience from it. Dalit aesthetics and literature pose an alternative to this paradigm of aesthetics, art, and literature. Responding to such traditions, Dalit literature, as exemplified by the select stories in this chapter, foregrounds an oppositional aesthetics that brings experience, expressions, and affect into the horizon of art and literature. Its approach is clearly and deliberately rooted in the socio-political, historical, and cultural context of Dalit life. Nature is always presented in its entanglement with life and labour. Its language is 'earthy' in the most material terms. This 'situatedness' in the ecological, socio-political and historical factors characterises the aesthetics of Dalit literature. In this way, Dalit aesthetics poses an alternative way of appreciating not just literature but all socio-cultural productions. Conceptualising ecology from a Dalit aesthetic perspective inherently integrates the themes of struggle and survival. Embracing Dalit aesthetics acknowledges how caste-based exclusion shapes their interactions with the environment.

References

Badiou, Alain. 2004. *Handbook of Inaesthetics*. Stanford University Press.

Bresler, Liora. 2006. Toward Connectedness: Aesthetically based Research. *Studies in Art Education* 48(1), 52–69. https://eric.ed.gov/?id=EJ767105.

Burke, Edmund. 1757. *A Philosophical Enquiry into the Origin of Our Ideas of the Sublime and Beautiful*. Printed for R. and J. Dodsley.

Currie, Adrian. 2023. Epistemic Engagement, Aesthetic Value, and Scientific Practice. *The British Journal for the Philosophy of Science* 74 (2), 313–334. https://www.journals.uchicago.edu/doi/abs/10.1086/714802.

Dangle, Arjun (ed.). 1992. Introduction. In *Poisoned Bread*, xi–xv. Orient Longman.

Dasan, M., Sanal Mohan, Pradeepan Pampirikunnu, Chuvappu (eds.). 2012. *The Oxford India Anthology of Malayalam Dalit Writing*, xiii–xxxiii. Oxford University Press.

Devy, Ganesh N. 2002. *Painted Words: An Anthology of Tribal Literature*. Penguin Books India.

Edachira, Manju. 2024. Anti-caste Aesthetics and Dalit Interventions in Indian Cinema. *Economic and Political Weekly* 55(38). https://www.epw.in/journal/2020/38/special-articles/anti-caste-aesthetics-and-dalit-interventions.html.

Foucault, Michel. 2014. *On the Government of the Living*. Palgrave Macmillan.

Kant, Immanuel. 1790. *Critique of Judgement*. Translated by J. H. Bernard. Kegan Paul, Trench & Co.

Kaplan, Stephen. 1987. Aesthetics, Affect, and Cognition: Environmental Preference from an Evolutionary Perspective. *Environment and Behavior* 19 (1): 3–32. https://doi.org/10.1177/0013916587191001.

Kumari, Priyanka, and Maninder Kapoor. 2021. Understanding Dalit Literature: A Critical Perspective Towards Dalit Aesthetics. *Rupkatha Journal on Interdisciplinary Studies in Humanities* 13 (4): 1–8.

Limbale, Sharankumar. 2004. *Towards an Aesthetics of Dalit Literature.* Orient Blackswan.
Madhukumar, M. K. 2012. Paalakkunnan's Journey. In *The Oxford India Anthology of Malayalam Dalit Writing*, eds. M. Dasan, Sanal Mohan, and Chuvappu Pradeepan Pampirikunnu, 81–84. Oxford University Press.
Mukherjee, Alok. 2004. Introduction. In *Towards an Aesthetics of Dalit Literature*, ed. Limbale, Sharankumar, Orient Blackswan.
Muthukkaruppan, Parthasarathi. 2017. Tamil Dalit Literature: Aesthetics, Politics, and Life. *Biography* 40 (1): 64–76.
Nayar, K. Pramod. 2015. Dalit Poetry and the Aesthetics of Traumatic Materialism. *Indian Journal of Gender Studies* 22 (1): 1–14.
Pambirikkunnu, Pradeepan. 2007. Adhasthitharude Utharaadhunikatha [The Downtrodden's Postmodernism]. *Mathrubhumi Weekly* May 13–19, 27–32.
Panicker, Ayyappa. 2022. The Aesthetics of Folk Literature. In *Ecocriticism in Malayalam*, ed. G. Madhusoodanan, 58–67. Cambridge Scholars.
Prasad, Indulata. 2022. Towards Dalit Ecologies. *Environment and Society: Advances in Research* 13 (1): 98–120.
Rancière, Jacques. 2004. *The Politics of Aesthetics: The Distribution of the Sensible.* Translated by Gabriel Rockhill: Continuum.
Rokade, L. S. 1992. To Be or Not To Be Born. In *Poisoned Bread*, ed. Arjun Dangle, 23–26. Orient Longman.
Vaduthala, T. K. C. 2012. Sweet-offering at Chankranthy. In *The Oxford India Anthology of Malayalam Dalit Writing*, eds. M. Dasan, Sanal Mohan, and Chuvappu Pradeepan Pampirikunnu, 53–60. Oxford University Press.

Dhanesh Mankulam is an Assistant Professor and Ph.D. Programme Coordinator at the Department of English and Cultural Studies, School of Arts and Humanities, CHRIST [Deemed-to-be University], Bangalore, India. Mankulam holds a Ph.D. in English from Kannur University, Kerala. He also holds an MA and a B.Ed. Degree. Offering a diverse range of courses at both PG and UG levels, he actively contributes to the academic landscape via teaching and research. Having recently authored an article for an edited volume on 'Dalit Concepts and Terms,' Mankulam has also contributed to the Oxford Bibliographies. He has also translated articles on O V Vijayan to an edited volume to be published by Routledge. His most recent publication is an article in an Encyclopaedia on New Populism published by Springer. He has more than ten publications to his credit.

Chapter 4
Ecological Injustices and Dalit Narratives: A Reading of Hira Bansode's Select Poetry

Ashima Shrawan

Abstract The present chapter investigates the complex interplay between ecological dynamics and the Dalits, a historically marginalised community in India. As a result of their extreme reliance on natural resources and manual labour, especially in the agricultural sector, Dalits, who were previously referred to as "untouchables", face unique environmental difficulties. They are more susceptible to the negative effects of climate change and natural disasters as a result of their vulnerability. Dalit perspectives and ecological knowledge play a crucial role in maintaining the environment, but they are noticeably lacking from mainstream environmental activism and literature. The present chapter explores Hira Bansode's poems which portray experiences of Environmental Casteism, which is defined by environmental injustices, like exclusion from natural resources, increased vulnerability to waste and pollution, etc. This chapter sheds light on Hira Bansode's poems such as "Petition", "Sanskriti", "O Great Man", "Bosom Friend", and collection of poems *Phyriad*, in which Bansode doesn't merely interpret the suffering of marginalised communities; instead, represents a voice that resonates empathetically with those at the fringes. She challenges divisions, contending that these are not merely mythical or cultural expressions but manifestations that serve the interests of the dominant social classes and castes. The chapter also highlights the necessity to incorporate Dalit perspectives into environmental discourse and activism for a more inclusive and just society.

Keywords Dalit · Ecological inequality · Caste · Untouchability

A. Shrawan (✉)
Shri Bhagwandas Adarsh Sanskrit Mahavidyalaya, Haridwar, India
e-mail: 14ashima@gmail.com

S. Kochar and R. Singh (eds.), *Pollution and Poverty*,
https://doi.org/10.1007/978-981-95-6876-5_4

4.1 Introduction

The connection between caste and ecology has been deeply ingrained in India's diverse landscape since ancient times, and has evolved amidst the current trend of globalisation that tends to homogenise cultural aesthetics.

> The historical origin of nature-writing and its lineage can be traced back further in the Vedic civilization of India. In contemporary Indian English literature, the trend of nature writing focuses on the question of creating ecological consciousness and settling the dispute between science-culture enigma… (Ghatak 2021)

Building on this lineage, the ecocritical writing in this context attempts to bridge the gap between viewing nature solely through a scientific or intellectual lens by emphasising its connection to landscapes and real-world settings. It explores themes like ecological ethics through myths, legends, folklore, inherited environmental traditions, forced migration, loss of rights and livelihoods, the survival of ethnic communities, local cultural activism, and sometimes simply the romanticised image of untouched natural beauty. These narratives often portray rural and indigenous communities as sharing the suffering of nature's exploitation, while the urban elite or bourgeoisie are typically positioned as the ones wielding power and contributing to this exploitation. Ecocritical writing in India has taken on diverse forms, appearing in ancient texts like the *Rig Veda*, *Ramayana*, *Mahabharata*, and Kalidasa's Sanskrit poetry, as well as in the works of some Mughal authors, British explorers, and historians. However, it gained significant momentum in early and post-independence Indian literature, with contributions from writers such as Salim Ali, Arundhati Roy, Ruskin Bond, Raja Rao, M. Krishnan, R. K. Narayan, Kamala Markandaya, Kiran Desai, and Anita Desai. The environmental movement also saw strong engagement from activists like Medha Patkar and Sundar Lal Bahuguna, especially from the late 1960s onward—a period that also highlighted the substantial involvement of Indian women.

4.2 Dalit Environmental Thought and the Politics of Belonging

Environmental politics in India has long been a contentious issue, encompassing struggles for rights and identity across various ecological fronts. While marginalised voices such as women and indigenous groups are included in these discussions, they often remain marginalised within the dominant framework. The Dalit community's engagement with environmental discourse is complex, as they suffer from the changing environmental dynamics while also resisting for mutual sustenance. In arid regions, Dalits often risk their lives to replenish ponds and water sources. From villages to cities, and from temples to schools, caste-based symbols of contamination, impurity, and filth pervade places and spaces, conjuring up fears of the presence of Dalits. For Dalit communities, landscapes such as forests and rivers embody complex and often contradictory meanings—forests may serve as both sanctuaries and spaces of exclusion or danger, while rivers, instead of symbolising purity or life, can become sites of violence and marginalisation, such as their use for the disposal of bodies. Nature, entangled with fear, violence, horror, and adversity, renders the environmental experiences of Dalits distinct and challenging. For instance, Dalit landless agricultural labourers in Kerala express such memories of forests, animals, woodlands, and weather through their songs.

This lived experience finds theoretical resonance in the work of B. R. Ambedkar, who prioritised the human-nature relationship equity and highlighted the unequal access to natural resources, particularly among marginalised groups like the Dalits. In the realm of Indian Dalit literature, which saw a significant resurgence in the late twentieth century and especially in the 1990s, narratives often explore Dalit livelihood within ecological contexts through various literary forms such as autobiographies, memoirs, plays, songs, ballads, folklores, poetry, and self-narratives. The songs *"Bhimrao Aaye Re"* (Dr. Ambedkar Has Come)—Traditional Dalit Folk Song, *"Nahi Hai Samman"* (There Is No Respect)—Marathi Dalit Lok Shahiri Song, *"Zameen Wapas Karo"* (Return the Land)—Contemporary Protest Song, carry refrains like: "*Na pani mila, na zameen hamari*" ("We got no water, no land of our own"). Gail Omvedt (1997) was among the earliest anti-caste scholars to argue that Dalit relationships with ecology cannot be understood apart from the structures of caste. In her influential essay, "Why Dalits Dislike Environmentalists," she identified two ideological currents—the anti-caste movement and the environmental movement—as central to explaining Dalit disengagement from contemporary environmental activism in India. According to Omvedt, the tension between these movements reflects a deeper historical divide in Dalit and non-Dalit conceptions of nature, represented most clearly in the contrasting perspectives of B. R.Ambedkar and Mahatma Gandhi. Many Indian environmentalists draw inspiration from Gandhi's idealized vision of the self-sufficient village, where human life exists in harmony with the natural world. In contrast, anti-caste thinkers generally resonate with Ambedkar's critique of the Indian village as a site of entrenched inequality,where caste hierarchies are normalized and Dalits are systematically excluded from social, spatial, and ecological participation.

These works frequently employ a confessional first-person narrative style, delving into myths, rituals, and traditions, reflecting a literature of ecocriticism and ecojustice. Despite their experiences of discrimination and ostracisation, Dalit narratives often find solace, hope, and a sense of equity and equality in the natural world. As Dalit narratives grapple with the loss of identity and the assertion of self within hegemonic socio-political and cultural structures, they often find refuge and assimilation with nature.

For example, in her memoir, Bama, the first published Tamil-Indian Dalit writer, describes herself as "Karukku" (meaning palmyra lives, symbolising a double-edged sword challenging oppression), reflecting her experience of oppression within the Christian Missionary system as that of a caged bird. Her journey of recovering from this social trauma is likened to a "falcon soaring high in the skies". (ix) Manohar Mouli Biswas, a Namashudhra writer from Bengal, envisions himself as a hyacinth (neglected in oblivion) and titles his work "The Life and Death of Prisnika", conveying his painful experiences. Dr. Siddalingaiah, a Dalit writer from Karnataka, resides in a separate area within a village called Ooru Keri and eventually passes away there. His autobiography shares the same title, reflecting a life close to nature, secluded from the structured violence of society and urbanisation. Thus, nature and the indigenous lifestyle, along with birth rights over the ecosystem, provide Dalits with a sanctuary against the harshness of the oppressive class binary system.

Within this context, the concept of "Dalit ecologies" presents itself as a crucial framework to analyse caste-driven segregation, loss of land and resources, dehumanisation, exploitation of Dalit labour and environmental wisdom, and the unequal effects of the worldwide ecological crisis on Dalit populations in India. Influenced by the self-identification of civil rights activists in the United States as Black, the term gained traction in India during the 1970s as a means to assert the political activism and radicalism of marginalised groups previously labelled as "untouchable" within the Brahminical (also spelled Brahmanical) Hindu caste structure. Across centuries, the socioeconomic practice of untouchability has shaped Dalits' understanding of their natural surroundings and their capacity to sustain themselves, paralleling the experiences of Black communities facing ecological discrimination rooted in race. During the early 2000s, a multitude of Dalit intellectuals and proponents against caste discrimination, such as Kancha Ilaiah, Gopal Guru, Chandrabhan Prasad, and Goldy George, voiced apprehensions regarding the environmental and political directions taken by Indian environmentalism. They argued that environmental movements showed scant regard for the nationalist, caste-based, hegemonic social, and environmental frameworks prevailing in the country. This critique is echoed in the following extracts from Dalit songs, which vividly depict ecological discrimination:

> O black girl
> You reap the paddy field
> You reap everything with your sickle
> Come along! O come along with the wrath of Kali! (Das 2012)

This poem, for example, presents a symbolic invocation of a rural, marginalised woman—referred to as the "black girl"—whose identity embodies both socio-economic struggle and latent power. Her act of reaping the paddy field signifies the routine agricultural labour performed by women in agrarian societies. However, the repetition and expansion of "reaping everything with your sickle" transcends mere farming, suggesting a broader metaphor for dismantling oppressive structures. The sickle thus transforms into a revolutionary instrument. The climactic call to "come along with the wrath of Kali" invokes the fierce Hindu goddess, aligning the girl's suppressed energy with divine feminine rage. This interweaving of mythology and revolutionary rhetoric positions the rural woman not as a victim, but as a catalytic force of social transformation and resistance.

Consequently, this interplay between caste and nature poses a significant challenge to Indian environmentalism, which has traditionally overlooked such connections. As Gopal Guru insightfully observes, "Social ecology makes dirt and filth an existential companion of Dalits who are at the receiving end of condescending descriptions of the former across time and space.… In the social construction of ecology Dalits become dirt and dirt is them" (Das 2012).

Similarly, sociologist Gail Omvedt drew attention to water access and exclusion in 1997: "Much today is written about traditional water harvesting systems. One example of these is the *phad* system in Maharashtra, and it seems to be what all environmentalists say it was: ecologically sustainable and providing equal water access to all cultivators". What often remains unacknowledged is that access to water was restricted solely to cultivators. Dalits—who were socially barred from agricultural activities—were consequently denied water rights as well. This exclusion raises critical issues concerning equitable access to water, the traditional belief that land belongs to those who till it, and the continuation of exclusionary practices under contemporary capitalist power structures. However, the conspicuous omission of these injustices in most environmental narratives is deeply troubling.

The noted environmentalist Mukul Sharma's book *Caste and Nature: Dalits and Indian Environmental Politics* delves into the complex and inseparable relationship between caste and the environment in India. He examines how Dalit experiences and narratives consistently highlight the daily ecological hurdles they encounter and how Dalits articulate their environmental experiences and aspirations. Professor Sharma has discovered that nature, intertwined with fear, violence, horror, and hardship, shapes the distinctive environmental experiences of Dalits:

> Eco-casteism is a representation, and more specifically an upper-caste Brahmanical representation, of nature and environment. Caste frameworks simultaneously dominate and delete sounds and narratives of Dalit environmentalism. (Sharma 2017)

Environmental historians, intellectuals, civil society groups, and various social movements have, whether knowingly or unknowingly, shown complacency—and at times complicity—in reinforcing a dominant and homogenised environmental discourse shaped by eco-casteism. Landscapes evoke deep-rooted caste anxieties regarding labour, blood, and bondage (Sharma 2017). In arid regions, Dalits often risk their lives to replenish ponds and water resources. From villages to cities, and from temples to schools, caste-related metaphors of pollution, impurity, and filth pervade places and spaces, reinforcing imagined threats associated with the presence of Dalits.

Dr Sharma approaches Ambedkar's work through an environmental lens to explore why mainstream environmental movements have largely excluded him from their canon, even as they retrospectively celebrate figures like Mahatma Gandhi, Jawaharlal Nehru, and Indira Gandhi. He emphasises that Ambedkar had a profound understanding of how caste shaped people's relationship with nature:

> Drawing from his life experiences of exclusion and denial, Ambedkar knew that water and other elements lose their vitality to "pollution" as soon as a Dalit comes in contact with them... [he]had a threefold schema of nature that mirrored the marginalization of Dalits in society: an "external nature," "universal nature," and "social nature." (Ray 2022)

Ambedkar, as Sharma presents him, was a staunch modernist. Rather than viewing "community"—often seen as separate from capitalist systems—as a potential space of resistance against colonial modernity, Ambedkar understood it as the very structure that sustained caste segregation and inequality. In the latter part of the chapter, Sharma reinterprets Ambedkar's engagement with technology and modernity as tools for dismantling caste-based community structures. This reading offers a counterpoint to the technocratic ideal promoted by Pathak and the Sulabh movement, instead proposing an alternative model that rethinks the interplay between humanity, technology, and the natural world.

With the inspirational leadership of Dr Ambedkar and the enabling environment post-independence, there emerged a newly-educated aspirational Dalit middleclass in Indian cities. These educated and self-aware Dalit young men and women felt indignant about their exclusion from the mainstream culture as it seemed to deny their very humanity and pushed them to the periphery. The supposedly great religious and cultural traditions invariably appear as hierarchical, hegemonic and exclusivist from the Dalit perspective.

The chapter will shed light on Hira Bansode's poems, such as "Petition", "Sanskriti", "O Great Man", "Bosom Friend" and collection of poems *Phyriad*, in which Bansode doesn't merely interpret the suffering of marginalised communities; instead, she represents a voice that resonates empathetically with those at the fringes. She challenges divisions, contending that these are not merely mythical or cultural expressions but manifestations that serve the interests of the dominant social classes and castes. The chapter also highlights the necessity to incorporate Dalit perspectives into environmental discourse and activism for a more inclusive and just society.

4.3 Environmental Injustice in Hira Bansode's Poetry

Hira Bansode, a Marathi Dalit poet, defied entrenched caste stereotypes that permeated various caste groups, opting for poetry as a means to articulate her anguish and disbelief regarding such entrenched and futile divisions based on caste. Through her poetry, she unveiled the shared consciousness of a community whose voice had long been suppressed throughout history. Whether Dalits, tribals, or women, they have perennially existed on the fringes, relegated to the margins of society's recognition. While recent awareness among the masses and the proliferation of literature have somewhat mitigated this practice, eradicating it entirely from our socio-cultural landscape remains exceedingly challenging. Hira Bansode grapples with several research inquiries, including how caste hierarchies endure through the exploitation of nature, the roles of purity and pollution in this context, the underlying principles determining access and exclusion, the possibility of assigning a caste to water, and how caste dynamics shape irrigation networks within villages. Furthermore, she explores why and how caste and its associated culture influence concepts of pure and impure food, impacting our dietary choices and preferences. The chapter centres on the literary contributions of Hira Bansode, a poet from the Mahar caste.

In her poetry, she confronts the abhorrent notions surrounding caste, highlighting the atrocities and injustices suffered by her fellow Dalits and women. Through the narratives of Shabri and Draupadi, she portrays the enduring suppression of marginalised groups throughout history; her verses critique the role of nature as well as religious, cultural, and societal institutions that have historically marginalised them. Her poetry delves into themes of alienation, liberation from various forms of oppression, estrangement, the quest for identity, and dignity. Bansode's writing challenges oppressive doctrines and amplifies the voices of the marginalised, especially women, who are often likened to a drum beaten at both ends. She questions the exclusion of untouchables from temples, asking how a god who deems them untouchable can truly be theirs. She further asserts that if the god excludes them, the religion cannot be truly theirs either. In recounting her life experiences, Bansode tells how she was not allowed to enter the temple constructed by her father. "But as my turn came, and I was about to enter the temple, the priest shouted at me, caught me by the hair, and dragged me from the temple entrance, refusing to allow me to enter" (Novetzke 1993). She asked why he did this to her. He replied, "I was a Mahar and could not enter because my presence would defile the holiness of the place. He said he would beat me, or kill me if I tried to enter again" (Novetzke 1993).

Hira entered the world as a Kamble, belonging to the untouchable Mahar caste. Despite living in a diverse neighbourhood, her social limitations were defined early on in her life. She recounts how Marathas barred them from entering their homes or even touching their water at the well. If a Maratha child interacted with them, punishment followed. "We could not enter the house of a Maratha, or touch their water when they came to fill at the well. If any Maratha child played with one of us, he or she would be beaten, yet as children, we were not fully aware of the concept of caste" (Novetzke 1993).

Her statement no longer shocks readers who understand how deeply caste segregation is rooted in the social fabric of many communities. Such occurrences are not just historical—they may still be ongoing. The unjust treatment and perception of marginalised castes and sub-castes, often justified by core Hindu religious texts, highlighted the urgent need for basic human rights that were systematically denied to them by the upper castes. Such occurrences, familiar to those aware of the entrenched caste divisions in society, reflect a historical and possibly ongoing reality. These actions, rooted in discriminatory beliefs upheld by texts of Hindu dharma, underscored the urgent necessity for basic human rights denied to lower castes by the upper echelons. Hira Bansode's poetry emerged as a vehicle for challenging entrenched norms that silenced Dalit and other oppressed voices, including women.

4.3.1 Sanskriti

"Sanskriti" by Hira Bansode is a poignant exploration of the environmental atrocities faced by Dalits, offering a critical lens through which to examine the intersection of caste oppression and ecological degradation. The poem delves into the profound implications of caste-based discrimination on the environment and the lives of marginalised communities.

> Great culture of this land,
> To hell with you!
> We are ashamed
> To call you, mother.

The opening lines appear to acknowledge the grandeur of Indian culture. However, they set the stage for a critical examination, suggesting that the so-called greatness is superficial. The poet challenges the oppressive aspects of the culture that marginalise Dalits and women, indicating that such a culture is not worthy of reverence. These lines express a profound sense of betrayal and disillusionment with the dominant cultural narratives of India, particularly those rooted in Brahminical patriarchy.

> As we burned beneath the summer sun
> You offered us no shade.
> Our humanity tested again and again.

The imagery "burned beneath the summer sun" underscores the inseparable link between social injustice and environmental degradation. The poem vividly portrays the lived experiences of Dalits, who are not only subjected to systemic oppression but also disproportionately bear the brunt of environmental injustices. Bansode highlights how the marginalised status relegates the Dalits to the fringes of society and she confronts environmental injustices in her poem. These injustices not only degrade the physical environment but also perpetuate cycles of poverty and marginalisation among Dalit communities. Through vivid imagery and evocative language, the poem challenges readers to recognise the interconnectedness of social justice and environmental sustainability. By centring the experiences of Dalit communities, "Sanskriti" urges readers to reflect on the structural inequities that underpin environmental degradation and to advocate for transformative change. Dr. Tejaswini Nandkumar Deo, a distinguished academician who has contributed significantly to marginalised voices in literature, also supports this perspective, stating, "Such vocal rejection is also an assertion of identity as a newly created social, religious and political community. The confident self-assertion is also a recurring feature of Hira Bansode's poetry" (Tejaswani 2022).

4.3.2 Petition

> The sky betrays as well
> With its heavy black clouds
> The Earth an accessory
> To these pre-arranged plots….

The lines, "The sky betrays as well, with its heavy black clouds, The Earth an accessory, to these pre-arranged plots", set the tone for a sombre exploration of the natural world as complicit in perpetuating systemic injustices. Here, the sky and the earth are depicted as collaborators in a predetermined narrative of oppression, symbolising the pervasive influence of caste hierarchy on the environment.

> …I petition
> Against the hidebound culture
> Imprisoning us in a cold cell,
> Giving us outcast lives….

The poet's petition "Against the hidebound culture, imprisoning us in a cold cell" underscores the entrenchment of caste-based discrimination within societal structures, which not only restricts individual freedoms but also stifles collective progress. The portrayal of Dalits as confined within a "cold cell" speaks to the dehumanising effects of caste oppression, which denies them agency and autonomy over their lives and environment.

> …This culture whose wind treats us like strangers
> In whose soil our seeds die,
> Whose rain provides us nothing but famine
> We reject this polluted life.

The poet laments, "This culture whose wind treats us like strangers, in whose soil our seeds die, whose rain provides us nothing but famine", drawing attention to the ways in which Dalits are marginalised and excluded from the natural world. The reference to seeds dying in the soil and rain bringing famine symbolises the systemic barriers that prevent Dalits from accessing resources and opportunities for sustenance and growth. In rejecting "this polluted life", the poet affirms the inherent dignity and resilience of Dalit communities in the face of adversity. By rejecting the status quo and advocating for social and environmental justice, the poem serves as a call to action for transformative change. It challenges readers to confront their complicity in perpetuating caste-based discrimination and environmental degradation, urging them to work towards a more equitable and sustainable future for all.

4.3.3 O Great Man

Bansode also incorporates socio-political occurrences to frame her poems. "O Great Man", translated by Jayant Karve and Eleanor Zelliot, pays homage to Ambedkar and his endeavours in advocating for Dalit rights. The poet makes a reference to the 1927 Mahad riot, during which Ambedkar led his followers to the *Chawdar Tank* to access its water. Drawing a comparison, Bansode likens him to an elephant, a significant emblem in the Dalit movement, symbolising the community's resilience and power. She talks about the environmental degradation faced by Dalits:

> It is clear that nature belongs to all
> but these people bought that too.
> Every drop of water in Chawdar Tank
> was stamped with their name…

The opening lines, "It is clear that nature belongs to all but these people bought that too", set the tone for a critique of the commodification and privatisation of natural resources, which disproportionately affects Dalits and other marginalised groups. The reference to "Chawdar Tank", a historically significant water body, underscores the symbolic and practical implications of the appropriation of natural resources by dominant castes.

> …the alert watchman of this culture
> guarded the imprisoned water.
> They feared that your touch
> would poison the water….

The imagery of water, a fundamental element of life, being "stamped with their name" and guarded by the "alert watchman of this culture" vividly portrays the extent to which Dalits are excluded from accessing essential resources. The fear that Dalit touch would "poison the water" reflects the deeply ingrained caste prejudices that underpin environmental injustices, perpetuating cycles of marginalisation and oppression.

The poem's critique extends beyond the physical realm to encompass the cultural and spiritual dimensions of environmental degradation. The desecration of Dalit lives and dignity is symbolised by the act of anointing them "with your blood when you were dying of thirst", highlighting the callous disregard for human suffering and the dehumanising effects of caste-based discrimination.

4.3.4 Bosom Friend

> You know in my childhood we didn't even have milk for tea much less yoghurt or buttermilk My mother cooked on sawdust
> she brought from the lumberyard

wiping away the smoke from her eyes....

This poem offers a poignant reflection on the intersection of caste, environment, and food scarcity, highlighting the profound socio-economic challenges faced by marginalised communities. The speaker vividly portrays the harsh realities of their childhood, characterised by deprivation and limited access to basic necessities, including food. The mention of not having milk for tea, yoghurt, or buttermilk underscores the economic hardships experienced by the speaker's family. These food items, often taken for granted by many, represent not only nutritional sustenance but also cultural and social markers of well-being and comfort. Dr R.P. Singh, founder-director of Food & Biotech Engineers, believes that milk and dairy products hold a vital place in the daily diet of people across the Indian subcontinent, deeply rooted in both cultural practices and nutritional habits. He states that milk is not merely a food in India; it is a symbol of purity, nourishment, and ritual significance, woven into the very fabric of daily life. (Singh 2019). The absence of such staples reflects the systemic inequalities and barriers faced by marginalised communities, including Dalits, in accessing essential resources. The image of the mother cooking on sawdust from the lumberyard, amidst the smoke irritating her eyes, encapsulates the resourcefulness and resilience required to navigate adversity. It speaks to the ingenuity and perseverance of individuals striving to provide for their families despite limited means and challenging circumstances.

...Dear Friend – Shrikhand was not even a word in our vocabulary
My nose had never smelled the fragrance of ghee
My tongue had never tasted halva basundi....

The absence of words like *Shrikhand*, the fragrance of ghee, or the taste of *halva basundi* in the speaker's vocabulary reflects not only a lack of exposure to certain foods but also a broader cultural and socio-economic divide. It highlights the disparity in experiences and opportunities between the speaker and their more privileged counterparts, whose access to culinary delights is intertwined with social status and economic privilege. The poem offers a poignant commentary on the complex interplay of caste, environment, and food scarcity, highlighting the resilience and resourcefulness of marginalised communities in the face of adversity. It prompts critical reflection on systemic inequalities and the urgent need for social and economic justice in ensuring equitable access to essential resources and opportunities for all.

4.4 Conclusion

Hira Bansode's personal struggle has found a place in her poetry; both the titles and content of her poems are piercing through the socio-cultural conclaves that have structured our society. Bansode explores a diverse range of themes in her poetry, placing equal importance on the empowerment of Dalits and women. Bansode's writings go beyond mere interpretations of suffering; they resonate empathetically with those marginalised in society. She critiques societal divisions, rejecting the notion that they are merely mythical or cultural, but rather constructs that serve the interests of dominant classes and castes. Her writing creates a space for those who have been overlooked, urging them to strive for equality in an uneven societal landscape.

In conclusion, the issue of Environmental Casteism faced by Dalits represents a complex intersection of environmental injustice and caste-based discrimination. Across various historical and contemporary contexts, Dalits have been systematically marginalised and denied access to basic environmental resources and protections. This form of discrimination manifests in various ways, including restricted access to clean water, land, and sustainable livelihoods, as well as exposure to hazardous environmental conditions and pollution.

Addressing Environmental Casteism requires a multifaceted approach that acknowledges and dismantles the deep-rooted structures of caste-based oppression within environmental policies and practices. It necessitates the recognition of Dalits' rights to equitable access to resources and decision-making processes related to environmental management and conservation. Moreover, combating Environmental Casteism entails challenging social norms and attitudes that perpetuate caste-based discrimination and advocating for inclusive and intersectional approaches to environmental justice.

By recognising and addressing the unique challenges faced by Dalits in accessing and benefiting from environmental resources and protections, society can move towards a more just and sustainable future for all. It is imperative for policymakers, activists, and communities to work together to dismantle the barriers of Environmental Casteism and strive towards environmental justice that is inclusive, equitable, and respectful of human dignity.

The entire ecosystem surrounding us consists of interconnected entities each distinct yet interdependent, comprising living organisms including animals and humans. Together, these beings rely on symbiotic relationships to maintain environmental harmony. However, in the current era of pervasive neo-capitalist socio-economic dynamics, societal divisions and inequalities are intensifying. In recent decades, India has experienced numerous social movements focusing on ecological concerns. However, like other political movements, these movements often overlook the marginalised yet deeply compassionate segment of the Indian population concerning ecological sustenance. The narratives explored thus far do not view natural elements as mere backdrop; rather, they reflect a profound philosophy rooted in Dalit's hierarchical yet simple, traditional, and minimalist lifestyles. The dominant and constructivist approach of contemporary movements has struggled to grasp the underlying exploitation of marginalised groups and their rightful connection to nature. Consequently, progress has been achieved, but it has predominantly favoured those in power. To break free from the grip of capitalist development and address this dual oppression we must reengage with confessional autobiographical narratives. By intertwining issues of identity with environmental equilibrium, these narratives can promote both existential resource usage and preservation as well as genuine developmental pursuits aimed at fostering a more equitable, diverse, humane, and progressive society.

Instead of disregarding caste, developing an awareness of its presence in nature can help establish a new political space for environmental struggles. Examining Indian environmental politics through the lens of Dalits and their myriad movements nationwide unveils a novel ecological framework—a discernible Dalit environmental public sphere—that often exists outside the prevailing discourse. Dalit environmentalism represents an ongoing and evolving initiative. Nevertheless, its limited visibility stems from the dominance of upper-caste figures entrenched in the sphere of secular modernity and citizenship. To enhance comprehension of caste, Dalits, and environmentalism, it is imperative to investigate diverse perspectives and interactions with the environment. Shedding light on concealed facets of secular environmentalism and its ramifications facilitates a deeper understanding of the intricate dynamics surrounding caste, Dalits, and environmental advocacy.

References

Bama. 2012. *Karukku.* Translated by Lakshmi Holmström. Oxford University Press.

Biswas, Manohar Mouli. 2011. *The Life and Death of Prisnika: Memoirs of a Dalit.* Samya.

Das, Ajay. 2012. O Black Girl. In *The Oxford India Anthology of Malayalam Dalit Writing*, eds. M. Dasan, S. Mohan, P. Pampirikunnu, and Chuvappu, 81–82. Oxford University Press.

Ghatak, Sourav. 2021. Nature-Writing in Indian English Literature: Ecological Consciousness and Cultural Aesthetics. *International Journal of English Literature and Social Sciences* 6 (2): 45–52.

Guru, Gopal. 2012. Social Ecology Makes Dirt and Filth an Existential Companion of Dalits. In *The Oxford India Anthology of Malayalam Dalit Writing*, eds. M. Dasan, S. Mohan, P. Pampirikunnu, and Chuvappu, 85–86. Oxford University Press.

Novetzke, Christian. 1993. Dalit: The Poetry of Hira Bansode. *Journal of South Asian Literature* 28 (1–2): 279–295.

Omvedt, Gail. 1997. Water Harvesting, Phad Systems and Dalit Exclusion. *The Hindu*, August 10.

Ray, Shashank. 2022. Ambedkar's Environmental Vision: Caste, Modernity and Nature. *Economic and Political Weekly* 57 (23): 35–42.

Sharma, Mukul. 2017. *Caste and Nature: Dalits and Indian Environmental Politics.* Oxford University Press.

Singh, R. P. 2019. *Handbook for Process, Evaporation & Drying Technology.* Food & Biotech Engineers India Pvt. Ltd.

Ashima Shrawan is an Assistant Professor of English at Sri Bhagwandas Adarsh Sanskrit Mahavidyalaya, Haridwar. Her main research area is Indian and Western Poetics. Her Ph.D. thesis has been published in the form of a book by Cambridge Scholars Publishing, United Kingdom. She has delivered a number of lectures, both at home and abroad, including Manchester and Nepal. She has published articles and papers in various journals, including Journal of Indian Philosophy, IUP Journal of English Studies, Indian Literature and Metacritic Journal for Comparative Studies and Theory. One of her papers was shortlisted for the prestigious C.D. Narasimhaiah Prize at the IACLALS Annual Conference in 2024.

Chapter 5
Roots of Exclusion: Examining Environmental Injustices in Balbir Madhopuri's *Against the Night* and Lal Singh Dil's *Poet of the Revolution*

Navdeep Singh

Abstract This book chapter examines the largely unexplored realm of Environmental Casteism as depicted in the selected Punjabi Dalit autobiographies, namely Balbir Madhopuri's *Against the Night* (2010) and Lal Singh Dil's *Poet of the Revolution* (2012). Through a critical lens, the chapter aims to unravel the relation between caste dynamics and environmental practices, exposing how Dalits, historically marginalised in the socio-cultural hierarchy, have been systemically denied access to natural resources. Drawing upon vivid instances from Madhopuri's and Dil's autobiographies, the study looks at specific incidents that underscore the prevalence of Environmental Casteism. One such example is the discriminatory prohibition against Dalits planting 'Aam' trees, a privilege reserved solely for the higher castes. The chapter also investigates instances where Dalits were denied the right to swim in the same water bodies as their higher caste counterparts, reflecting an egregious form of environmental exclusion based on caste identity. Instances of denying Dalits the right to cultivate specific trees and restricting their access to common water sources are proofs of the entwined dynamics of caste-based discrimination and environmental injustice. The analysis employs an interdisciplinary approach to unravel the deeper implications of these instances within the broader socio-cultural context. By examining the personal narratives of Balbir Madhopuri and Lal Singh Dil, the chapter aims to contribute to an in-depth understanding of how Environmental Casteism intersects with the daily lives and struggles of Dalits.

Keywords Caste · Casteism · Dalit · Dalit autobiography · Environmental Casteism

---

N. Singh (✉)
M. R. Govt. College, Fazilka, Punjab, India
e-mail: navdeeppanjawa3@gmail.com

S. Kochar and R. Singh (eds.), *Pollution and Poverty*,
https://doi.org/10.1007/978-981-95-6876-5_5

5.1 Introduction

Environmental justice has emerged as a critical area of research, focusing on the inequitable distribution of environmental benefits and burdens across different social groups. However, the intersection of environmental injustice with caste-based discrimination within the context of Punjabi Dalit communities remains a relatively underexplored area. This chapter addresses this gap by investigating the manifestations of Environmental Casteism as depicted in Balbir Madhopuri's *Changiya Rukh: Against the Night* (2010) and Lal Singh Dil's *Poet of the Revolution* (2012). These autobiographical narratives open a window into the lived experiences of Dalits in Punjab, displaying the systemic denial of access to natural resources and the broader implications of environmental exclusion. Through a critical examination of these autobiographies, this chapter seeks to display the specific ways in which caste dynamics influence environmental practices and contribute to the marginalization of Dalit communities.

Balbir Madhopuri and Lal Singh Dil are two of Punjab's most significant Dalit literary voices, whose works provide insights into the caste fabric of Punjab. Balbir Madhopuri, born in 1955, is an eminent Punjabi writer and translator. His autobiography, *Changiya Rukh: Against the Night*, is considered to be the first Punjabi Dalit autobiography to appear in English. His autobiography narrates his journey as a Dalit in Punjab, where he has to face caste-based discrimination. Similarly, Lal Singh Dil (1943–2007), a major revolutionary Punjabi poet and activist, uses his poetry and prose to highlight the experiences of being a 'Chamar' in Punjab. His autobiographical work, *Dastaan* or *Poet of the Revolution: The Memoirs and Poems of Lal Singh Dil*, remains an important work to study caste oppression and the socio-political struggles of Dalits in Punjab. Through their narratives, both authors provide firsthand accounts of marginalization and the enduring caste-based exclusion from natural resources.

In India, the caste system has been a defining feature of its social fabric for centuries, which categorises individuals into hierarchical groups called *varnas* with varying degrees of privilege and disadvantage. This deeply entrenched system affects nearly every aspect of life, including access to natural resources and exposure to environmental hazards. Higher castes typically enjoy greater access to clean water, fertile land, and other natural resources, while the lower castes, especially Dalits, face systemic exclusion from these essential environmental benefits. Caste is thus a critical factor in environmental discrimination. Lower caste groups are disproportionately deprived of environmental benefits and bear the brunt of pollution. Despite the crucial role caste plays in environmental injustice, "the environmental discourses in India usually ignore or do not give adequate consideration to this aspect," as observed by the Dalit scholar Bhimraj M. in his influential work "A Dalit Critique of Environmental Justice in India" (2020).

Historically excluded from having equal access to natural resources, Dalits are often relegated to segregated, unclean living spaces, unhygienic working conditions, and restricted from accessing water, shade, even soil, due to purity-pollution taboos. Contemporary Dalit scholars Vijay Kumar and Binod Mishra conceptualise this exclusion through the term "casteisation of natural resources," arguing that "for Dalits, common pooled resources such as water, land and forests, schools and colleges, socio-cultural spaces and various other public institutions have become a symbol of caste oppression and exploitation" (2022). This "casteisation" extends beyond rural landscapes, permeating urban settings, where Dalits are similarly confined to "dirty" occupations and marginalised spaces.

The caste dynamics of Punjab is shaped by a complex interplay of historical, religious, and socio-economic factors. While Punjab is predominantly Sikh, a religion founded on principles of equality, caste remains a persistent reality. Punjabi sociologist Paramjit S. Judge explains that due to the Sikh ideology and its emphasis on equality in worship, "the Sikhs opened their religious places for the lower castes" (2015). However, despite these egalitarian ideals, caste inequality persists within the community. Judge further explains, "Towards the end of the 19th century, the Sikh ideologues began to argue that Sikhism did not believe in caste," yet the continued presence of caste-based discrimination suggests these claims remain unfulfilled (2015). The state has a significant Dalit population, comprising nearly one-third of its total population, with communities such as the Mazhabi Sikhs, Ravidasis, and Chamars forming the largest Dalit groups. Despite formal Sikh doctrines promoting inclusivity, caste-based segregation is evident in villages, gurudwaras, and land ownership patterns. Dalits, often landless, work as agricultural labourers or in menial jobs, with limited access to resources like water and land. Upper-caste Jat Sikhs dominate the agrarian economy and political landscape, further entrenching disparities. The resilience of caste in Punjab underscores the contradiction between religious egalitarianism and social realities, making the region a critical site for studying caste-based environmental injustices.

The ecological symbols of land and water, central to both *Against the Night* and *Poet of the Revolution* reflect the socio-ecological exploitation of Dalits under the caste system. As Mukul Sharma, who has written extensively on caste and ecology, insightfully observes, "earth, land, water, agriculture, and forest are not just spaces or places; they are also bound by structures of inclusion and exclusion based on touch, feel, and food" (2017). Through a close reading of these autobiographical narratives, this chapter brings forth the stark realities of Environmental Casteism in Punjab to show how ecological exclusion reinforces and sustains the marginalization of Dalits.

5.2 Land: The Scarred Earth of Caste

Talking about Punjab's unique socio-cultural landscape, Ronki Ram, a well-known political sociologist and a scholar of Dalit studies, points out, "what distinguished Punjab from the rest of the country is that caste inequity persists here more in terms of landownership, social identification and dominant cultural patterns than of Brahminical orthodoxy" (2020). Harish Puri, an eminent scholar of Punjab's social history, also comments, in the "Introduction" to *Against the Night*, "[t]he history of the oppression of dalits in Punjab was related more to the structure of economy, ownership of land, and the politics of power relations, rather than the Brahmanical worldview" (Madhopuri 2010). The Jat community's monopoly over agricultural land, contrasted with the widespread landlessness among Dalits, shows the stark social hierarchy in the region. Ronki Ram further discusses, "Since Punjab happened to be primarily an agricultural state, the ownership of land assumed significant importance in determining social status" (2020). This adds to the marginalization of Dalit communities who are deprived of this essential resource.

In *Against the Night*, Balbir Madhopuri discusses land ownership and environmental injustice, showcasing the harsh realities faced by Dalit communities in Punjab. Land, as portrayed in this personal narrative, is not merely a physical entity but a symbol of power, social status, and the deep-rooted casteism that pervades the agrarian society of Punjab. In the "Introduction" to *Poet of the Revolution*, the prominent Indian poet and translator Nirupama Dutt observes:

> Punjab has a higher percentage of Dalit population than other states. Scheduled Castes form about 30 per cent of the total population and 8 per cent of these caste live in the rural areas and are landless. The landowners are mostly Sikh Jats. (Dil 2012)

Through their control over land, the Jats assert their superiority by limiting the Dalits' access to vital resources. Madhopuri's narrative vividly illustrates how Dalits are pushed to the peripheries of villages, both literally and metaphorically. Ronki Ram also describes the setting of Punjabi villages "are divided into upper caste and Dalit settlements" and that these "settlements are located, invariably, on the side towards which the dirt of the village flowed. Dalits were not allowed to build pucca (concrete) houses because the land on which they lived did not belong to them (2020). Similarly, in the "Introduction," Madhopuri's day-to-day-village life is described as:

> The two segments of the living space were clearly marked: *Pind* (the village) on one hand, and *Vehra* or *Chamarli*, the segregated row of dalit houses located on the lower western periphery of the village, on the other. (Madhopuri 2010)

Here, the symbolic naming of land based on caste deepens this discrimination. The way how the settlement where "Chamars" reside is normally called "Chamarli" without realising how casteist a land becomes this way. In another instance, Madhopuri highlights that the place where Punjab Singh Ramadasia, a Chamar, sits to make shoes, is commonly referred to as "gandian da chappar" (the pond of the cobblers). These names reflect a social order that not only physically but also symbolically marginalises Dalits. Also, the spatial segregation extends beyond residential areas to include wells, temples, and cremation grounds, which further divide the worlds of the upper castes and Dalits. Madhopuri explains this phenomenon by stating "settlements of the untouchables are always in the lower ends…this is because this class of people, in accordance with the social system, are not part of the caste system and do not belong within the four varnas; even their shadows are to be avoided (2010).

He further adds:

> Though it is asserted that they are within the Hindu fold, actually this statement is meant to keep the dalits permanently enslaved. That is the why they have been kept out of the mainstream. (Madhopuri 2010)

In contrast, Madhopuri remarks, "Within a few years of Baba Sangatia's settling here, pucca houses came up in the area. As a result, the future generations were called *pucca wale*—of pucca houses—and the place from where sand was quarried is still called *gora chappar*—the fair pond" (2010). This contrast underscores the link between caste identity and access to durable, permanent housing, which serves as a marker of social status. However, despite these small gains, the systemic inequities continue to persist. Madhopuri mentions, "According to the official records of the patwari, this is common land and the lower castes still have the right to take mud from this place for plastering their mud-huts… But the Jats are gradually taking over these lands" (Madhopuri 2010). Despite being designated for use by lower castes, common lands such as these are gradually being encroached upon by the dominant Jats, depriving Dalits of even the minimal resources available to them.

Madhopuri records, in his village, "The common land is 17 acres and this includes the lanes and ponds of the village. There are thirty wells in the field. But how much land do the untouchables have? Almost nothing!" (2010). This resonates with when Ravikumar, a leading Tamil critic and Dalit thinker, indicates, "A 'common well' means one from which an untouchable cannot draw water, a 'common funeral ground' means a place where the body of an untouchable cannot be cremated …," in the same way 'common land' means one that does not belong to a Dalit (2009).

Madhopuri's reflections on land ownership laws further expose the systemic injustices faced by Dalits. He notes, "During the hundred years of British rule in Punjab, the Punjab Land Revenue Act, 1887, remained in force, which prevented the untouchables from buying land even when they had the money" (2010).

Although some reforms were introduced post-independence, such as the East Punjab Land (And Prevention of Fragmentation) Holding Act. Under this act, "every low-caste family in a village was given two *marla* of land for the dumping of household waste" (Madhopuri 2010). Additionally, "the right of *marusi* [the common land] was eventually abrogated in 1957 and the Punjab Land Revenue Act (Intiqale Arazi Act), 1887, which prevented dalits from buying land had already lapsed, and the scheduled castes now had the right to buy and sell land" (Madhopuri 2010). However, by this time, the damage had already been done—Dalits were left with little to no land to call their own.

Against the Night delves into the everyday indignities suffered by Dalits, who are subjected to physical and psychological violence by the upper castes. Madhopuri recounts, "If a low-caste boy were to come out on the lanes of the village, all bathed and dressed in new clothes, his hair combed, one or the other of the Jats sitting under the trees would get up and throw mud on him. If he protested, he was sure to be beaten up" (2010). This showcases the broader societal norm that sought to keep Dalits in a state of perpetual subjugation, where even their aspirations for a clean and better life were met with hostility.

This narrative captures the hostility and violence that Dalits face when they attempt to challenge these norms. In one instance, a zamindar (landowner) threatens, "If I don't pulp all these Chamars with my boots, I am not the son of a Jat!... They've cut the fodder from my field under the pretext of cutting weeds and grass! As if the field is their father's!" (Madhopuri 2010). Here, the violent assertion of land ownership and the derogatory reference to the Dalits as 'Chamars' show how small they are considered in the caste hierarchy that dictates who has the right to use the land.

Phumman, Madhopuri's uncle's son, articulates the frustration and helplessness of the Dalit community, "No one says anything to those who harass us day and night! Do we not belong here that they have made life so difficult for us? They scold and terrorise us all the time" (Madhopuri 2010). While talking about the importance of land, he claims,

> Merely because they have lands, they have become our gods! They are the ones who feed us! If we did not work so hard in your fields, you would starve! You will forget to drink and raise hell in the evening, the way you do! (Madhopuri, 2010)

This persistent tension between the Jats and Dalits encapsulates the broader conflict within Punjab's agrarian society. "The Jat-Dalit conflicts thus signify the emerging Dalit assertion and its serious implications for the asymmetrically structured agrarian society of Punjab," as observed by Ronki Ram (2020). Furthermore, Phumman's words show how the "caste has divided environmentalism into 'full-stomach' and 'empty-belly' environmentalism, where the latter stands merely for the subsistence of the poor" (Kumar and Mishra 2022). This shows for Dalits, the struggle for basic survival takes precedence over broader environmental concerns, as the Dalits work tirelessly on land they do not own, only to be exploited and oppressed by those who do. Caste hierarchies have rendered most Dalits landless, even though the majority work as agricultural labourers in rural areas (Bhimraj 2020). A critical instance of this exclusion is exemplified in Madhopuri's reflection on his father's experiences:

> I had seen my father storing the grain in the granaries and bins at Iqbal Singh's house, and heard him mutter, 'Today, we walk barefoot on this mound of grain, but once it is stored, we would not be allowed to touch it, our touch contaminates it.' (Madhopuri 2010)

This statement reveals a deep-seated prejudice where the physical labour of Dalits is essential for the collection and storage of grain, but their touch is deemed contaminating once the grain is stored. This exclusion shows how Dalits are exploited for their hard work while being denied basic rights and respect.

Illustrating the deeply ingrained Environmental Casteism in Punjab, Madhopuri recounts an incident where he excitedly takes out a mango sapling to plant in the courtyard of his home. This simple act of planting a tree is abruptly interrupted when Bhaia, his father, snatches the hoe from his hand. Bhaia admonishes him, saying, "Maama, you are trying to ape the Jats! They have large lands and large havelis! We only have this much space where we can sit and relax" (Madhopuri 2010). This interaction reflects a deeper social reality where the ownership and cultivation of land, particularly the planting of trees like mangoes, are acts laden with cultural and caste-based significance. This prohibition serves to reinforce the social hierarchy, where

Dalits are expected to remain in a state of subservience, even in their interaction with the environment.

Madhopuri's narrative deepens this sense of deprivation through Bhaia's lamentation over their lack of land. Bhaia would often speak to himself, saying, "If only we had a small piece of land, we would have been very comfortable. No one knows which bastard has deprived us of land" (Madhopuri 2010). He further exclaims, "If I ever come to meet him, I shall tell that saala, that he should work like a slave the way we do, and if his bum doesn't burst, then he may come and catch me!" (Madhopuri 2010). Watching Bhaia's despair, Madhopuri dreams of a future where he could own land, plant orchards of bananas and mangoes, and grow roses near the well, much like the prosperous Jats. This dream, however, remains out of reach, highlighting the systemic barriers that prevent Dalits from owning and cultivating land. Madhopuri's vision of planting orchards and flowers symbolises a longing for connection to the land—a connection that is systemically denied to Dalits through entrenched caste-based practices.

The Dalit experience of environmental injustice is also evident in their interactions with the landowners, as Madhopuri recalls:

> My father worked for different zamindars, and we would go to the houses of these zamindars, sit in their courtyards, and put out our bowls for rotis. Their women threw the rotis from the top and we would adroitly catch them. (2010)

To avoid the touch, "The dal and vegetables were also ladled out from above into our bowls, and sometimes blistering hot splashes fell on our feet" (Madhopuri 2010). This dehumanising practice of feeding Dalit labourers from a distance further shows the caste-based exclusion that permeates all aspects of life, including the most basic human interactions. Madhopuri's father's longing for land is also poignantly expressed when he says, "If only we had a few acres of land, then we would have not bothered about these mean zamindars!" (Madhopuri 2010). This desire for land ownership represents not only a quest for economic stability but also a profound yearning for dignity and social equality.

In *Poet of the Revolution*, Dil reflects, "My mother's younger brother used to work as a *siri*, or landless labourer, for the landlords. This amounted to becoming a slave…" (2012). This highlights the oppressive and exploitative nature of the *siri* system, where landless labourers, often from lower caste communities, worked for landlords under conditions that were akin to slavery. This *siri* system traps individuals in a cycle of poverty and servitude, which perpetuates their socio-economic vulnerability. In another incident, Dil recounts how policemen would brutally assault him, taunting, "So you bloody Chamars, you want our land?" (2012). This reveals the rooted caste-based oppression faced by Dalits, where the desire for land was curtailed by brutal resistance and dehumanization by those in power.

The systemic exclusion of Dalits from land ownership and their relegation to environmentally degraded areas are indicative of the broader power dynamics at play in Punjab's agrarian society. As Indulata Prasad, a scholar of Dalit studies, points out, "Land marks material, political, and symbolic privilege while modulating caste-laced idioms of inequality, dominance, and subordination" (2022). The narrative's exploration of land and environmental injustice serves as a powerful critique of the socio-economic structures that continue to oppress Dalit communities.

5.3 Water: Dividing the Flow

Water, an essential element of life, becomes a tool of oppression and exclusion in the context of caste hierarchies in Indian societies. Madhopuri's *Against the Night* offers a stark depiction of how water, much like land, is a site where Environmental Casteism manifests. As Bhimraj M. emphasises, "Landlessness either by unequal distribution or deprivation and locating untouchables in the outskirts far from the main habitat are the main factors which handicap them to utilize the resources" (2020). The geography of the village itself is a testament to these deep-rooted injustices. The western part of the village, where the Dalits reside, is deliberately positioned lower, ensuring that all the dirty water flows in their direction. As Madhopuri explains, "it is believed that not only should they not pollute clean water but also that these people should live in dirt, mire, and slime" (2010). This physical segregation forces the Dalits to live in perpetual filth. Madhopuri vividly describes this "hateful, inhuman system" as one that continues to plague villages across India, underscoring how environmental degradation is imposed upon Dalit communities as a form of social control.

The division of water sources further exemplifies this exclusion. As Madhopuri recounts, "The village appeared to be one, but every group had its own drinking wells" (2010). This segregation is not merely a matter of convenience but a deliberate act of keeping Dalits away from shared resources. Even when they had access to water, it came at a cost. Madhopuri recalls how he often saw "Tai Sibo and her son, Jeet, both of the Jhiwar caste, fetching water for various households," a task relegated to their caste only (2010). This labour, essential for the functioning of the village, was performed by those who were simultaneously deemed unworthy of sharing the same water sources as the upper castes.

In another incident, Diwan, a member of the Jhiwar caste, expresses the internalised acceptance of caste roles by scolding his son for attempting to deviate from their traditional duty of fetching water: "I say, why are you making a spectacle of yourself? This is our work, the work of the Jhiwar caste" (Madhopuri 2010). In stark contrast, Diwan's younger son, Pashu, expresses the anguish and rebellion of the younger generation, who are unwilling to accept the dehumanising and exhausting labour imposed on them by their caste status. Pashu's outburst, "This is no bloody life! It is so difficult! Winter and summer… fetch water for people first thing in the morning, and only then can we go to school! From today, I am not going to school" (Madhopuri 2010). This accentuates the physical and emotional toll of this labour. The harsh realities of fetching water before even attending school reveal the daily struggles faced by Dalit children, whose lives are marked by servitude and limited opportunities.

The indignity associated with water use is further highlighted through the interactions between caste groups. Madhopuri narrates how Jat children would wash the tap thoroughly after a Dalit child had used it. He narrates, "I had often noticed that whenever we used the tap, any Jat child who used it after us would wash the tap thoroughly before drinking from it. This often made me wonder what was wrong with my hands that others had to wash a tap after I had touched it" (2010). This act of washing the tap after its use by a Dalit child is emblematic of the entrenched belief in the pollution of touch, a belief that permeates all aspects of life, including something as fundamental as access to water.

The social hierarchy extends to the very act of drawing water, with caste determining who can access which water sources and how they must do so. For instance, Madhopuri's Bhaia recounts an incident he was threatened by Inder Singh, a Jat, for drinking directly from the pipe:

> He told me to advise my brother-in-law (Gulzari Lal) not to drink straight from the pipe carrying water from the Persian wheel, but only from a lower level, or else he'd teach him a lesson. (Madhopuri 2010)

Furthermore, as Madhopuri recounts, once after finishing their work, they approached a hand pump, only to be stopped by Sodhi's wife: "'Stop, I'll give you water to drink.' She made a sign and we went to the back near the gutters, and cupped our hands, as she poured water from a jug, held above our heads" (2010). This demeaning act of making Dalits drink water in a manner that avoids any physical contact between them and the water source starkly exemplifies the social ostracism imposed upon them.

Additionally, "the two-wheeled well stood on the eastern edge of the village, and only the women of Jat, Brahmin, and goldsmith communities went to this well with their offering of sawain" (Madhopuri 2010). The well then is not merely a source of water but a site where caste-based rituals are performed, further entrenching the social hierarchy. This shows how water access is tightly regulated and controlled, with any perceived transgression met with hostility and violence. For Dalits, the denial of access to this well represents yet another layer of environmental and social exclusion, one that perpetuates their marginalised status within the village.

A similar dynamic of caste-based control over water resources is reflected in the experience shared by Lal Singh Dil in *Poet of the Revolution*. Dil recalls an incident during a literary gathering, stating, "Once, Kanwal, a celebrated leftist novelist, was invited for a literary gathering. We had all sat down for a meal. I was reaching out for a jug of water when Kanwal pulled the jug away as though only, he had the right to drink that water" (2012). This incident shows the persistence of caste discrimination even in intellectual and progressive circles. It features the painful reality that caste-based exclusion extends into all facets of life, regardless of the setting.

Even in death, the caste-based segregation of water persists. In Madhopuri's *Against the Night*, the issue of water injustices extends beyond the living and into the rituals surrounding death. The segregation and inequality that Dalits face in life persist even after death, as seen in the narrative of a flooded cremation ground. Madhopuri recalls how, after a woman dies at night, her cremation must be delayed due to the flooding of the cremation ground. However, the eastern side, which is higher and thus susceptible to flooding, is reserved for the cotton corders' community, while the Adidharmis, to which Madhopuri belongs, have access only to the waterlogged western part (Madhopuri 2010). This becomes evident when Joginder from the cotton corders' community points out, "That part of the cremation ground is for our use" (Madhopuri 2010). This emphasises the rigid boundaries that continue to separate different Dalit sub-castes even in death. Madhopuri describes the scene as:

> [T]he cremation ground for the dalits was common to all, but the path to Rastgo passed through it, dividing it into two. On the basis of population, the larger portion of the west was accepted as belonging to the Adidharmis, and the eastern part to the cotton corders. (Madhopuri 2010)

Nonetheless, Madhopuri adds, "over a period of time, this distinction had been all but forgotten. Yet, now and then, tradition reared its head. It had now" (2010). This incident underscores the deep-rooted caste biases that dictate every aspect of life and death in the village. The fact that even the cremation ground is divided along caste lines, with certain areas deemed more suitable or accessible than others, reflects the exclusion of certain Dalit communities.

The internalised caste hierarchies among Dalits themselves are also evident in Madhopuri's narrative's another anecdote. A Dalit character reflects on the division within the Dalit community, saying,

> The Jats are men of substance and landowners. Just look at these Ramdasias! They are our people, and merely because they have accepted Sikhism, they act superior to us! At one time they drew water from our wells, but now they have their separate well! (Madhopuri 2010)

This internal caste division further depicts the pervasive nature of caste discrimination that infiltrates every aspect of life, including access to water. He further adds, "They say that the land near and around our wells reeks of tobacco from the water of the hookahs! They don't say it outright, but they are no longer the same as us!" (Madhopuri 2010). Madhopuri's account is also filled with poignant memories of how Dalits were treated worse than animals when it came to water. He recalls a scene where a herd of 'Ramgai' drinks at a pond in which the water fell from the Persian wheel, a sight that pains him. Madhopuri asserts, "It hurt me to think that we were considered worse than animals! Even stones and dumb animals are accepted and considered better than us! Or at least considered more valuable and important! They are even worshipped" (2010).

In *Poet of the Revolution*, Dil recounts an episode from his childhood, when he went swimming:

> Once, for instance, following a Brahmin who resembled Tagore, I entered the little pool of water by the well where the Jat boys were bathing. I must have then been in either class II or III. I was wearing a dirty shirt and torn drawers. (2021)

However, he describes the aftermath of such a doing as, "The water drawn from the water was flowing in a torrent from a sprout and I had just crouched down to put my head under the water when a Jat boy pulled me out and gave me a thrashing" (2012). So, the lower caste people were not allowed to enter a space reserved for higher caste even if it is a little pool. He acknowledges, "my mother had to apologise to his mother for my dreadful mistake of entering the sacred space of the upper caste" (2012). The fact that his mother had to apologise for his "mistake" further emphasises the social norms that perpetuated this systemic injustice.

These narratives show how deeply caste hierarchies are embedded in the social fabric of rural Punjab. Water, a fundamental resource, becomes a powerful symbol of marginalization and exclusion, where Dalits are denied equal access to it.

5.4 Tree: Roots of Exclusion

Madhopuri's autobiography is originally titled *Changiya Rukh* in Punjabi, meaning "a tree lopped from the top, slashed and dwarfed" (Madhopuri, 2010). The title itself powerfully reflects the systemic exclusion that has robbed Dalits of opportunities for growth and prosperity. In the chapter "The Banyan Tree of the Chamars," Madhopuri draws upon the metaphor of a lopped tree to encapsulate the truncated lives and potential of the Dalit community, an 'untouchable' caste "whose potential for growth has been 'robbed by the Hindu social order'" (2010).

The banyan tree stands as a central symbol of the community, under which the Chamars of the village would gather. Despite their marginalised status, the Chamars managed to buy a small plot of land—"the sixteen marla (484 square yards) of land under the bargad" from Kartar Singh of Neevan Vehra (Madhopuri 2010). As Bhaia recounts, "The whole transaction was by word of mouth, and no document was signed. Bhai Harbans Singh of the Puccawalas himself measured the area" (Madhopuri 2010).

Dalits have had a long history of social and environmental struggles, where they frequently clash with Brahmin and *savarna* domination of eco-space. As Bhaia further mentions, "Those days the law did not permit Churas and Chamars to own land, or build houses, let alone for cultivation, even if they could pay" (Madhopuri 2010). These struggles have also led to the "assertion and creation of an autonomous Dalit eco-space," which challenge the structures of caste-based exclusion (Sharma 2017). In this context, Madhopuri's narrative brings to light how the banyan tree in Chamarli, the segregated Dalit section of the village, functions as a vital eco-space for the Chamars. This tree was not just a part of the physical landscape but a witness to the community's lives, sheltering them during monsoons. Not just that, but "it was also a witness to many happy events, bitterness, tensions, and fights" (Madhopuri 2010). This tree symbolised a place of gathering, celebration, and mourning—a central piece of the community's social fabric. As Madhopuri writes that the "area of the looms under the trees was clean and open, and therefore something or other was always going on here" and that it was the place where "marriages were fixed and weddings were held. The baraats were put up here. The area served as an assembly ground on occasions of bereavement" (2010).

The memories attached to the banyan tree are vivid and significant in Madhopuri's autobiography. For instance, the sants from Mehuna once organised an Akhand path of the Darbar Sahib under the banyan tree near the looms. The tree was also the place where Rulia, the barber, cut the hair of the community's children, and where swings were set up on the lower branches for the children to play. Madhopuri describes the importance of the bargad tree as: "Our mohalla was the center of entertainment for the whole village because of the bargad trees" (2010). Thus, this banyan tree was more than just a landmark; it was the epicenter of social and cultural life in the Dalit community. He further adds:

> It was here that mimicry shows were held, and Saal was organised—an annual feature, it was organised after the monsoon, when everyone prayed for the welfare of cattle. It was considered a part of the duties of the Chamars. (2010)

The tree served as a communal space where celebrations, rituals, and even everyday interactions took place, reinforcing a sense of community and belonging. However, even the tree could not escape caste-based discrimination that permeated village life, as "[t]he members of other castes, such as Jats, Brahmins, carpenters, barbers, kahars referred to the tree as the bargad of the Chamars… [and] Jats would use expletives for the tree" (Madhopuri 2010).

The eventual cutting down of the banyan tree in February 1972 marked a traumatic and symbolic act of violence against the Dalit community. Madhopuri recalls returning home from school to find "[t]he bruised and massacred twin trees were an unbearable sight" (2010). The felling of the banyan tree was not just a physical loss but a symbolic one as well, representing the erasure of the Dalit community's history, culture, and resilience. Madhopuri's words capture the deep sense of loss:

> Looking at the broken and battered branches on the ground I felt as if the protective shade on our heads had been torn asunder. The loom pits seemed naked and vulnerable. The sole witness to all the oppression we had endured had been felled and destroyed. (2010)

Madhopuri further contextualises this loss within the broader environmental and socio-political changes taking place in Punjab. He reflects on how, "[i]n the last five or six years, the situation in this land of five rivers had deteriorated sharply… Men seemed to have turned into monsters and were doing their best to control even nature" (2010). Yet, he holds onto a glimmer of hope, believing that "[o]ne day nature will surely be kind to man. The earth and the sky stay in place because there is justice and righteousness" (2010).

Through "The banyan tree of Chamars," Madhopuri underscores the profound connection between the Dalit community and the natural environment, a connection that was brutally severed by the forces of casteism. The banyan tree, once a symbol of communal life and resilience, became a casualty of the systemic oppression that sought to keep Dalits in a state of perpetual subjugation. Through these vivid accounts, Madhopuri and Dil put forth the harsh realities of Environmental Casteism.

5.5 Conclusion

The exploration of Environmental Casteism in Madhopuri's *Changiya Rukh* and Dil's *Poet of the Revolution* shows the deeply entrenched systems of oppression that have historically marginalised Dalits, both in terms of social hierarchy and environmental justice. This chapter looks at how land, water, and even trees—essential elements of life—become sites of social exclusion and discrimination.

The chapter shows that the environmental injustices faced by Dalits are not merely ecological concerns but are deeply intertwined with social and cultural oppression. The exclusion from land, water, and other environmental resources reflects the pervasive nature of caste-based discrimination. As the chapter reflects on these narratives, it becomes clear that the fight for environmental justice is inseparable from the broader struggle for social equality.

References

Dil, Lal Singh. 2012. *Poet of the Revolution: The Memoirs and Poems of Lal Singh Dil*, trans. Nirupama Dutt. Delhi: Penguin Books

Judge, Paramjit S. 2015. Caste Hierarchy, Dominance, and Change in Punjab. *Sociological Bulletin* 64: 55–76.

Kumar, Vijay, and Binod Mishra. 2022. Environmental Casteism and the Democratisation of Natural Resources: Reimagining Dalit Testimonies. *South Asia: Journal of South Asian Studies* 45: 577–595.

Bhimraj, M. 2020. A Dalit Critique of Environmental Justice in India. In *Contemporary Environmental Concerns: Multi-Disciplinary Aspects of Environmental Law*, ed. Renuka Soni, 89–120. Punjab: Rajiv Gandhi National University of Law

Madhopuri, Balbir. 2010. *Changiya Rukh: Against the Night*, trans. Tripti Jain. Delhi: Oxford University Press

Prasad, Indulata. 2022. Toward Dalit Ecologies. *Environment and Society* 13: 98–120.

Ram, Ronki. 2020. Burden of Past and Vision of Equality: Political Sociology of Social Exclusion and Jat-Delhi Conflicts in Punjab. *Desh Doaba*. http://www.deshdoaba.com/burden-of-past-and-vision-of-equality-political-sociology-of-social-exclusion-and-jat-dalit-conflicts-in-punjab/. Accessed 17 June 2024.

Ravikumar, A. 2009. *Venomous Touch: Notes on Caste, Culture and Politics*, trans. R. Azhagarasan. Kolkata: Samya

Sharma, Mukul. 2017. *Caste and Nature: Dalits and Indian Environmental Politics*. Delhi: Oxford University Press India.

Navdeep Singh is a PhD Research Scholar in the Department of English and Cultural Studies, Panjab University, Chandigarh. He is currently working as Assistant Professor at M.R. Government College, Fazilka, Punjab. His research interest lies in the areas of Dalit Studies and Translation Studies.

Chapter 6
Affective Land and Dystopian Environmental Desire in Raju K Vasu's *Polappatham*

Krishnan Unni P

Abstract In the light of the existing debates on Dalit writing and environment discourses, one is struck with the possibilities of engaging with and exploring the conditions of temporalities that operate in Dalit conditions. In Malayalam Dalit writing, this has raised a number of theoretical issues such as locating the Dalit narratives and the past (s) they try to recontextualise, the perpetual difference seen in the temporal conditions of Dalit lives, and the ontological and ethical status of Dalits punctured in an environment of living and forgetting. This chapter tries to look at Raju K Vasu's novel *Polappatham* to situate three basic themes connected to Dalit environments and narratives. The second area of analysis is the historical contingencies of Dalit lives and their breach with the different conditions of environment that makes the Dalit the 'Other' to certain other communities. The third area of exploration is the connectivity of Dalit environment and the narration, where their ontological status is both a matter of difference and non-static political representation. Above all, this study argues for a Dalit environment that would challenge the Dalit essentialism by inching Vasu's narrative to some of the crucial questions of the narrative deviance and Dalit expressions.

Keywords Environment · Community · Affect · Alterity · Difference

6.1 Introduction

The environmental conditions of the lower castes in India are the balance sheets of the discarded places from where they were either ousted or forced to live. In postcolonial India, such a situation of the Dalits invites a discussion of three important aspects: the first and foremost is the double marginalisation the Dalits face with respect to their status and dignity in various places. The second is the historical conditions of their past. The third is the unobtrusive political condition of the Dalits, where their

K. Unni P (✉)
Department of English, Deshbandhu College, University of Delhi, New Delhi, India
e-mail: apskup@yahoo.co.in

S. Kochar and R. Singh (eds.), *Pollution and Poverty*,
https://doi.org/10.1007/978-981-95-6876-5_6

struggles in the past are nostalgically and spuriously reflected in our times. The sally back into the past of the Dalit lives does not merely signify their environmental crisis of contemporary times. It necessitates, on the other hand, a deep link of the colonial past with some elements of the postcolonial conditions. The seamless continuity and the occasional breaks between the pre-colonial times and colonial period have relegated the Dalits to further margins; where they had to rearticulate themselves regarding their ways of coping with the environment. Scholarly discussions of the Dalit environmentalism often take for granted that it is deeply entwined with the set of practices and institutions related to the localities. Various discourses of the caste environmentalism, needless to say, have situated the overarching questions and doubts of mainstream environmentalism away from the daily activities of the Dalits. This has resulted in the double marginalisation of the Dalits both in their public and personal lives. Nishant Ghatak in an interesting article "Caste, Ecology and Nature": Environmentalism in Indian Dalit Literature "traces the ecosystem" of the Dalits by formulating Ambedkar's views: "the monolithic and constructionist paradigm of the present ideologies has been failing so far to understand the evil roots of exploitation on these outcastes and their subsequent rights over nature for a long time. As a result, progress has been made but is mostly compromised under the echelon of power" (2021, p. 111). Among the various reasons that segregate the Dalit ecology from the mainstream, the most important is the pre-ordained notion that they always have a fixed space/place in nature. In this regard, the discursive idea of environment becomes a tool for the privileged communities to elude accountability of their social privileges towards the marginalised, which subsequently silences the voice of the subaltern, specifically the Dalits and thereby disempower them, rather than recognising Dalit voice as an impetus to create a broader public awareness.

As Indian environment is diverse and deviant on account of multiple geographical structures, demography and ecological variance, Dalit lives are predicated not on any fixed assumption of ecology. The varied and multiple constituents that function in the castes are related to a number of socio-political features such as territories and non-distinctions, exodus of the Dalits from place to place, the marginalisation of the Dalits from natural resources, urban vs rural divisions and concomitant pressures exercised on them by several power structures. Environmental Casteism often subverts the condition of the Dalits based on particular locales, and hence, it creates a mimetic fantasy of mapping many issues of the lower castes within certain totalising systems. The Dalit lives in Keralam offer a special case study as the lower castes have been successfully woven into the fabric of misrecognition and misappropriation. As complex and capacious variants of perpetual existence over the decades, several castes in Kerala have been undergoing territorial demarcation and double marginalisation not only from the upper castes; but from sub-castes, which tend to sideline inter-caste environmental issues. On the one hand, there were attempts to study caste as a homology in Kerala by several sociologists; but on the other hand, the discontinuities among the castes and their temporal existence augmented casteism more powerfully. The constitution of different temporalities of castes in Keralam, it must be noted, doubled the internal caste divisions and ecological balance of people. The Dalit writers of Kerala have approached this issue from various points of view, of which the most important, is a journey back to the source of the pertinent lower caste. Of the recent Dalit writing in Kerala, the most prominent ones that look at the temporal discontinuities created by various power structures are the narratives of Johny Miranda, V. R. Santhosh, and Raju K Vasu. While the first two writers are deeply concerned with depicting the environmental issues as metaphors, Raju K Vasu's narratives offer a different stance by offering multiple registers of pre-colonial, colonial and postcolonial histories to weave strands of Dalit differences. This chapter looks at Raju K Vasu's novel *Polappatham* to develop a different discourse of Environmental Casteism—that is, not just the one recreated by upper castes and other groups of people; but by a set of differences created by the Dalits both internally and externally in the habitat of their existence.

6.2 Mapping the Initial Coordinates of Environment

Though since centuries, the Dalits in Keralam have been living in different spaces/places—both urban and rural, most importantly, their rural existence, is still a matter of grave concern. In the sociological caste structure of Keralam, a few castes stand different due to their social and economic conditions. The Enlightenment spearheaded by social reformers such as Narayana Guru, Chattambi Swamikal, Ayyankali, Sahodaran Ayyappan, and others helped in developing a new caste consciousness in Keralam from the beginning of the Twentieth century, which resulted in the official recognition of the lower caste people in the state. As the first state of India, where the people elected their government through the ballot box in 1956, Keralam claims to have a special place in the political history of the country, where its people, at an earlier stage, were concerned about the exploitation of the lower castes. The reservation of the lower castes in the assembly of Keralam and in many job sectors have shown a difference by highlighting their dignity. However, these coordinates of the lower castes, never could address the questions of the Dalit spaces, where, the lower castes underwent deep humiliation and contempt. The space is the most important coordinate of Dalit environmentalism. The figuring of the Kerala Dalits in historiography, therefore, is a matter of heavy contention and incrimination.

As social reformers gave thrust on the upliftment of the Dalits from their social conditions by offering different perspectives on different ways of looking at their environmental conditions, the community consciousness was centred on certain fixed cultural statuses and practices of the selected Dalit communities with respect to the environment. For example, Parayas and Pulayas—two suppressed castes in Keralam, had different caveats of community sensibilities since time immemorial. Their practices were not only different, but deeply marred by the ecological conditions that overwhelmed their daily course of lives. Relegated to the margins of villages, small forests and river banks, Parayas never found an opportunity to interact with the other socially depressed castes in Keralam. Though there were unofficial interactions between different socially oppressed castes, such interactions were under scrutiny of different people, who belonged to both the upper and lower classes. Environmental

Casteism, hence, depended not just on the assimilation of the lower caste people into the mainstream; on the other hand, its dependency rested on the ways in which the demographic and ecological discontinuities where castes were deeply entrenched. As the environmental topology is wide and varied, it raises a series of questions before us: can there be a Dalit narrative from Keralam showcasing Environmental Casteism can be projected as a form of uniform literature? In what ways can we unwind the so-called assumptions of the hegemonic practices that mark the Dalit narratives non-global? What further steps can be taken by Dalit writers and thinkers to mark Environmental Casteism an issue of global importance, and not just one of regional status?

Raju K Vasu's fictions such as *Chaavuthullal* and *Polappatham* have opened new ways of understanding Environmental Casteism based on the ways in which caste historiography is constructed from the perspective of official history writing and they vehemently stress the need of a cultural caste narrative, which would elucidate the environment as a desire and contingent feature. In all his narratives, Vasu tries to create a liminal space not in the postcolonial sense of "periphery", where the subalterns and the oppressed reside; on the other hand, as a space that has the potential to coexist with the provincial power centres. Looking at the existence of the social space, Sanal Mohan writes: "From the perspective of Dalit history it is difficult to think of civil society without referring to the mobilisation against caste slavery in the mid nineteenth century. The social space and civil society in Kerala were caste determined. The modern middle class that became decisive in the formation of civil society was almost exclusively constituted of the upper castes – including the Ezhavas, although they are referred to as a backward caste" (2008, pp. 74–75). Mohan's observation makes it clear that certain privileged castes were also determining the space of Keralam since the nineteenth century. Vasu's primary attempt is to rip open the space of differential environmental polarities. This has been elaborated further in the context of Vasu's *Polappatham* after theorising the environmental space.

David Harvey theorises nature and caste in the line of ecological equanimity, democratic unity and perpetual existence by systematising all ecological projects "political and social" (1996, p. 71). The idea Harvey proposes is "the environment of justice" (p. 81). However, Harvey's thrust is based on the notion of linking various inhabitants of nature to ecology by positing justice as the key factor of equanimity. The ways of looking at the capitalist mode of production by infringing the space of the natural inhabitants do not figure in the study of justice by Harvey, though such attempts are varied in different countries. Mukul Sharma, on the other hand, powerfully offers a critique of ecological capitalism and the method in which new food and sustenance are provided to the people of nature in global times. Sharma states that "… in a post liberalization world, a caste perspective of environment justice seeks to foreground the knowledge of the working of caste structures amidst new economies, power, science and technology, and their role in subjugating people behind a rhetoric of development" (2023, p. 45). Sharma's argument is valid when we understand the myriad ways of caste oppression in India based on environment. However, the "Caste capital-economy" Sharma argues in his studies requires a more historical and political foundation as in the Indian context, we come across a wide array of castes distanced from nature due to political compulsions. Both the above theorisations of caste take us to the fundamental ways of tracing the past of environmental conditions that date back to the pre-colonial and colonial times. Environmental Casteism in global times is intertwined with pre-colonial and colonial past, both in their own ways were responsible for formulating the postcolonial condition of ruptures.

Set against the backdrop of the late nineteenth and the early twentieth centuries, *Polappatham* charts out the lives of the Pulayas, Parayas, Suriyani Christians and other groups of people in the mid-part of Keralam, where the missionary education played a crucial role in molding the lives of the lower castes. Vasu's novel is replete with a lot of caste practices that were practised since time immemorial in Keralam by the Parayas and Pulayas that went unheeded despite the colonial conditions in India. Polappatham is a structure made by woods on top of which banana strands are kept to induce the bad demon in the Pulaya community. Structured as a discourse between Chinnayi who hails from Paraya caste and Ulaki from the Pulaya, the text conflates the internal mechanism of caste as both castes have a number of issues that make them distinct. The Pulayas are more oriented towards the hard work on the agricultural fields and other cultural practices such as invoking the spirits. The Paraya, on the other hand, are more inclined towards the archetypes and art forms, and they are known traditionally for weaving mats and rugs. The cultural practices of both castes are deeply embedded in various environments. Koluvamooppan, who is known as the living god of the Parayas is one such person who keeps agile the traditional worship practices of the caste. We read in Vasu's *Polappatham*, one of the ancient methods of setting the lamp: "Mooppan will light the lamp by standing on the step above. Wind from the east will augment that. Mooppan should lift the lamp to his shoulders" (2021, p. 41).

Traditional practices such as the above were unmarred by the colonial infringement. The birth of Makaran, the son of Kolavan, who travels into the hidden regions of jungles is yet another instance where these practices are challenged. Marimuttu, an adivasi from the jungle warns Kolavan that his son should never set his foot in the jungle. By bringing the polarities between the jungle dwellers and Pulayas, the text brings into focus the decline of magic. Vasu writes: "People dependent on Kolavan for magic and medicine started decreasing in their numbers. When the two girls who got married came back with *kuzhikkanam*, Kolavan returned to the traditional job of weaving mats and rugs. He started feeding them by fish, *noonikka*, and *attikka*" (2021, p. 61). A number of similar references indicate confrontations in the beliefs of castes related to the question of their environmental adaptability. These primary coordinates, needless to say, function as the bedrock of Environmental Casteism as they are dependent on cultural practices of castes in Keralam.

Different communal practices and exchanges among the Parayas and Pulayas are the initial coordinates that marked their history before some of them underwent conversion to Christianity. Both in terms of labour productivity and beliefs, such practices enriched the environmental casteism not as a stream of conflicts, but as a necessity of going back to the roots. The roots of Environmental Casteism have been examined based on affect and alterity as *Polappatham* gives us conflicting registers of co-living of the lower castes.

6.3 Straddled Alterities and Affect

Attempting to read *Polappatham* as part of alterity of caste and an affective dynamic, the concept of 'roots of deviance' is proposed as a device of historical and political employment. Unlike certain spatial engagements we read in the narratives of Dalit Malayalam writers such as Johny Miranda, Narayan and others, who use the space as fable, metaphor and an imaginary trope, Vasu constructs the space as a mundane fixity surrounded by characters who construct their imaginary pasts. In the environmental history of India, alterity has a special significance, which is set in such a manner by which, the binaries fast disappear. Environmental history, as suggested by Michael H Fisher, depends on "increasingly created climate changing factors over recent centuries" (2018), which successfully replaced a number of histories of the tribes and other people. Ramachandra Guha, on the other hand, sidesteps the issues of the Dalits in his *Environmentalism: A Global History* as he tries to locate his analysis within the contexts of a set of records from Indian Vedas, classics to the efforts of Gandhi. Guha contends, "The actual experience of scientific forestry was quite often at odds with its professed aims and supposed achievements. Especially in the colonies, it followed a 'custodial' approach, with the strengthening of state control having as its corollary the denial of customary rise of user exercised by peasant and tribal communities" (2016, p. 38). In such attempts, what is essentially missed is the veritable temporalities at work as the lower castes of India are intricately related to environment since several centuries. Derek Attridge examines the inextricable relationship of the singular and the other in *The Singularity of Literature*. Attridge argues that "Otherness.....is produced in an *active* or *event-like* relation- we might prefer to call it a *relating*" (2004, p. 29). In postcolonial times, alterity has helped us to think beyond the binary divisions—I/you, us/they, and native/coloniser. Alterity achieves new dimension when Vasu portrays a set of characters such as Mathai *Thambran* (Lord Mathai), Susan and a set of other Christian characters who are related to the labour, education and later reformation of the lower castes in Keralam. As the postcolonial alterity implies the state of being or other, the condition of locating the 'other' of the lower castes in the text is a suspect.

Mathai *Thambran*'s past is not officially recorded in history. Though the arrival of the Christians in Kerala goes back to the fables and records of the arrival of St. Thomas in Kodungallur, Mathai *Thambran*, who is not associated with the missionaries, but a landlord who owned several Pulayas, is an antinomy in the Suriyani Christian history of Keralam. Vasu writes: "Nobody knows from where did the ancestors of Mathai came and how did he become a Christian" (2021, p. 98). In order to etch them in history, though much later, the members in the family of Mathai *Thambran* begin to write their family history. The unmarked and non-spoken histories of the lower castes, we read in the text: "Not even in the pages of the family history, Parayas and

Pulayas never found a space…but in the clutters of Pulayas and others, amid the chewed spits of their generations, their histories spoke" (2021, p. 99).

The relationship between the Christian landlords and the lower castes, Vasu suggests, at times accentuated the Environmental Casteism through the labour of the Dalits, and at other times, intensified the rifts between them. Hidden in this logic, works the construction of the 'Other.' Set against the usual standards of the thematic and ideological reason, the urge of Parayas and Pulayas to find the history of their landlords, never functions as a binary opposition. While the lower castes remain steadfastly trapped within the gap of the Christian missionary attempts of conversion and the innate thrust of going back to their roots, the anxiety generated functions as a strategy of making them subsume before the exploiters of their environment. This strategy is the deviant form of viewing the lives of the lower castes in Keralam as an oppressed space of humiliation. Vasu, on the other hand, provides us a 'space of alterity', which commingles the Suriyani Christians and the lower castes in a far symbiotic way, where the idyllic environmental life is portrayed with a tinge of longing. This alterity makes the masters and the subjects exist to alter the environment.

The imaginary and the symbolic that work along with the histories of the obfuscation of castes from their environment in the postcolonial context challenge the notion of land as an *apriori* space. It is argued that this condition is an 'altered affect'. Forests, villages and the narrow spaces where Parayas and Pulayas thrive in the text counter Henri Lefebvre's notion of the "representational space" (1991, p. 41), which locates the symbolic and imaginary "the sources of inconsistency and incoherence" (p. 42) and further as "historical constructs in a very concrete and perhaps immediate way: they embrace, in other words, 'the loci of passion, of action, and of lived situation': (p. 42)". Challenging Lefebvre's "representational space", Thomas Master and Susan take their journey to educate and reform the lower castes. As their journey is charted out from various regions to Kanjirapally, a place in Kottayam district, the flora and fauna of Thiruvithamkoor are much taken into consideration an altered form of affect as we have instances of the Christian missionaries and their evasive encounters with the caste. The lower castes in Keralam, as Vasu shows us, were far deviant from not just the upper caste people or their masters; but from their sub-castes and tribes. This brings us to the condition of affect—much theorised by recent socio-cultural critics such as Gilbert Simondon, Julia Kristeva and Brian Massumi. Affect is not just a category, but a condition that sets the Parayas and Pulayas different in *Polappatham*. Chinnayi's and Ulaki's doubts are centred on the doubts of physical interaction and transaction of different objects, which were derided by their ancestors since a long time. Affect is a mental condition in the history of caste and environment in Keralam, whose roots go back to the pre-colonial times. The efforts of Suriyani Christians to educate and reform the caste, needless to say, encouraged mutual interaction, but instilled the fear that the lower castes have to sacrifice their beliefs. The vestiges of pre-colonial times, Vasu portrays, offer the scope of a journey back to the source, where their systems were intact with the music of environment.

The history of the lower caste resistance against the upper castes in the first half of the Twentieth century by social reformers provides us a different dimension of affect and alterity. It must be noted that though Ayyankali, a reformer of the Pulaya caste and a great thinker, we read in *Polappatham*, never gives any thrust on the environment. The efforts of Ayyankali resulted in the recognition of Pulayas to claim the roads, which were long denied to them for travel by the upper castes. The shift from the locales to the public space, it should be noted, is an important step that opened a new wave of thinking. Ulaki's story of Kotha, a lower caste lady who worked as a maid servant of Egerton complicates affect. Kotha's suicide from jumping from the mountain and her escape from conversion, as Ulaki narrates in her story, was a jolt against Kathiran, her lover who is deeply entrenched in the customs and traditions of the caste. Her death is an escape from embracing a fixed affect—provided by the foreign priest. Descriptions such as "She flew like a bird to the evening clouds that were resting in the evening sun before Kathiran, who was straddled" (2021, p. 187) make us aware of the lower caste woman's final desire to be a part of the nature, from where she has come to be a part of the land, where she lived many years.

The theorising affect in *Polappatham* raises issues of environmental alterity. Extending the notion of affect from corporeality and elaborating that from the ideas of Gilles Deleuze and Timothy Murphy, Brian Massumi writes: "Affect is as good a general term as any for the interface between implicate and explicate order. Turning to the difference between the physical and the biological, it is clear that there can be no firm dividing line between them, nor between them and the human" (1995, p. 99). Massumi's observation demonstrates that the affect in the environmental issues of the lower castes is more than corporeal and psycho-centric; but at the same time, it establishes a tenuous relationship with the environment of the past lives of the Dalits. Seen in this light, the suicide of Kotha signifies a brave and powerful return back to the environment of her ancient times. The environmental alterity, it must be noted, is a step to diagnose the past of strangeness and wilderness. The consequent death of Kathiran by the stumping of an elephant after Kotha's autodafe marks the significant moment in the text. This environmental affect, needless to say, gave way for many Dalits in Kerala to later embrace Christianity. The maelstrom of affect produced a sudden and fervent embracing of ideas of separation and loneliness as a dominant aesthetic principle during the time of caste commingling. Looking at the condition of Dalits, who are derooted from their soil, Vinyl Paul writes: "Dalits started believing in Christianity at a time when they were fully denied of social justice and equality from the Christians" (2021, p. 94). The body of Kathiran, around which "a few elephants were hooting till people came" (2021, p. 194) signifies the belief in the law of the jungle. It is strongly believed by the sub-castes that if the search for the dead by the lover is not fulfilled, the jungle will take another toll. In other words, the journey back to the source and the modernisation by clearing the jungle are the poles apart aspects that constitute the environmental alterity of the caste.

6.4 Conclusion

Raju K Vasu's *Polappatham*, as argued in this chapter, is not just a text of Environmental Casteism that highlights essentialism. Environmental Casteism in this text resists the inclusive notion of linearity of the Dalit lives in Keralam. Such a totalising notion could only be fulfilled at the cost of generating further exclusions. Alterity and dystopia—the twin themes, this chapter has attempted to theorise, permanently contest the idea of the environment as a subject of a complete discourse, and generate the postcolonial paradigm of difference concerned with lower castes. The resonances of alterity and dystopia can be seen in *Chavuthullal* in the form of fear when Vasu writes: "She screamed when she saw the shape of something which stood out from her shadow." (2013, p. 198). The idea of "environmental universality" needs to be left permanently open, contingent, in order not to foreclose in advance any claims for exclusion. The Parayas and Pulayas and the partially appearing shadows of other castes in Keralam intimately constitute a habitual pattern of environment. As their histories were much distorted and efficiently sidelined, Vasu stresses on the cultural practices, ethnographic features and on the language of the castes to produce retrieval of their memories of environmental past.

Environmental Casteism in the context of Keralam is a bundle of internal contradictions, differences and the cultural practices of the lower castes, which as Vasu's text significantly highlights, is dependent on the distancing of castes from their own habitations and furthermore, what is sensitised through touch as a mechanism of affect. "Complicating the features of touch, Aniket Jaaware observes", the dalit's shadow for us is a metonymy, but it is not necessarily a metonymy for the orthodox practising brahman. "For him the touch of the shadow is the touch of the dalit" (Jaawaare 2019, p. 102). Jaaware's statement is absolutely valid and a slight extension of that would make us realise that the internal divisions of the castes and sub-castes in Keralam are generated out of this fear. When touch is translated into fear, as the adivasis, Parayas, and Pulayas of Keralam have encountered, that makes them deviant from the jungles, rivers and all other natural habitations. The pre-colonial past nostalgically recollected is fraught with both the colonial and internal colonial (Anglican Christians) interventions that accelerated the above deviance. The text, therefore, attempts to synthesise different temporalities to suggest that Environmental Casteism in Keralam is straddled with multiple tasks of realising the many pasts of the lower castes. In other words, such a task would enable us to retrieve the past through an imaginative cultural re-invocation.

References

Attridge, Derek. 2004. *The Singularity of Literature*. London: Routledge.

Fischer, Michael H. 2018. *An Environmental History of India: From Earliest Times to Twenty First Century*. London: Cambridge University Press.

Ghatak, Nishant. 2021. Caste, Ecology and Nature: Environmentalism in Indian Dalit Literature. *A Journal on Theory and Praxis* 2 (2): 100–112.
Guha, Ramachandra. 2016. *Environmentalism: A Global History*. Delhi: Penguin Random House.
Harvey, David. 1996. *Justice, Nature and the Geography of Difference*. London: Wiley-Blackwell.
Jaaware, Aniket. 2019. *Practicing Caste: On Touching and Not Touching*. Hyderabad: Orient Blackswan
Lefbvre, Henri. 1991. *The Production of Space*, trans. Donald Nicholson-Smith. Oxford: Blackwell.
Massumi, Brian. 1995. The Autonomy of Affect. *Cultural Critique* 3: 83–109.
Mohan, Sanal. 2008. Social Space, Civil Society and Dalit Agency in Twentieth-Century Kerala. In *Dalit Studies*, ed. S. Ramanarayan and K. Satyanarayana, 74–103. New Delhi: Permanent Black.
Paul, Vinyl. 2021. *Atima Keralathinte Adrishya Charitram*. Kottayam: D.C. Books.
Sharma, Mukul. 2023. Environmental Justice and Caste After Liberalisation. *Economic and Political Weekly* 58 (13): 1–10.
Vasu, Raju K. 2013. *Chavuthullal*. Kottayam: DC Books.
Vasu, Raju K. 2021. *Polappatham*. Kottayam: DC Books.

Krishnan Unni P is Professor in English at Deshbandhu College, Delhi University. He is a creative writer in Malayalam and in English and has published five books. He has edited Gabriel Garcia Marquez's *Chronicle of A Death Foretold* (World View Publications) and his debut novel in Malayalam, *Keralam: Oru Documenta* won the Karoor Neelakanta Pillai Award in 2016. His latest book is on the Italian philosopher and theorist Giorgio Agamben, published as part of the Theory series by S.P.C.S. Kottayam, Kerala. He is a member of the Cultural Studies group in Cairo University, Egypt and a representative of the Memory-Studies Network at University of Lund, Sweden. He has presented papers in the University of Nottingham, U.K, Cairo University (Egypt), University of Lund, Sweden, the National University of Brussels, and at several other universities and academies in India. He had been a board of Studies member in the Kashmir University and Nainital University, Kumaon and at Mar Athanasius College, Keralam. His areas of interests are the Third World literatures and films, gender formations, changing patterns of sexual dissidence, and the politics of the dispossessed—concerned with music, football, and popular culture.

Chapter 7
Ecology of Plight: Reading of Selected Bangla Dalit Short Stories

Umesh Patra and Shubhadeep Jana

Abstract The word *Dalit* connotes the *untouchables* and the downtrodden in Indian society. Their marginalisation pertains not only to their caste or their position in society, but extends to their access to ecology and space. This chapter explores how Dalits face ecological precarity in their everyday life. The short stories are taken from an anthology, *Survival and Other Stories: Bangla Dalit Fiction in Translation* (2012) edited by Sankar Prasad Singha and Indranil Acharya. The chapter studies "Survival," which depicts a father's battle with a wild beast for food, and "On Water and On Shore," detailing a fisherman's perilous sea voyage to make ends meet. It also examines "Farmer Gopal's Caste and Creed," highlighting Gopal's evolving identity amid ecological changes, and "Munnali," which portrays her occupation as a manual scavenger, leading to her eventual disillusionment. Such stories have the potential to exhibit how Dalits are dependent on nature and natural elements, and yet how their marginality makes them face ecological plights of different kinds. Ecologically, they are marginalised to the point where their livelihood is obtained either by rummaging through dirt or by fighting with wild animals. Loss of forests, loss of lands, and loss of their identity put them in a crisis. The analysis of these tales foregrounds the broadscale environmental injustice meted out to them by the upper caste people of the society.

Keywords Dalit · Ecology · Plight · Suffering

U. Patra
Department of English, School of Humanities and Languages, Mahatma Gandhi Central University, Motihari, India
e-mail: umeshpatra@mgcub.ac.in

S. Jana (✉)
Department of English, Presidency University, Kolkata, India
e-mail: shubhadeepjana98@gmail.com

S. Kochar and R. Singh (eds.), *Pollution and Poverty*,
https://doi.org/10.1007/978-981-95-6876-5_7

7.1 Introduction

The term *Dalit* serves today as an umbrella term for the communities that were historically marginalised on the basis of caste in India. Derived from Sanskrit, the word *Dalit* literally means *pounded* or *crushed*. Metaphorically, it connotes the people who were ostracised, discriminated against because of their birth in such caste groups in India, which were deemed lower in the caste-hierarchy. Dalit Literature is a thriving branch of Indian literature, written in multiple Indian languages, comprising multiple genres ranging from fiction, autobiography, poetry, non-fiction and others. Oral renditions and audio-visual performances of Dalit narratives have also proliferated. Responding to the question "What is Dalit Literature," Muktibodh (1994), a noted Dalit critic and author, says that "Dalit literature is the literature produced by Dalit consciousness" (p. 272). By and large, Dalit literary specimens have been objective in nature, aiming at social reconstruction in order to eventually establish a casteless Indian society.

Caste has been a glaring problem in Bengal. The dismissal of the complex interplay of caste question that is fraught with other socio-political and economic concerns can be viewed as a deliberate attempt to propagate the myth of a caste-less West Bengal. In colonial Bengal, Dalit movements were led by, mainly, two communities: Rajbanshi and Namasudra. The dominant discourse of such movements was carried forward by two popular Matua reformers Harichand Thakur (1812–1878) and his son, Guruchand Thakur (1846–1937). Matua is a Namasudra sect that followed Vaishnava traditions to eradicate the social hierarchy. *Matua Sahitya* as resistance literature emerged during the late nineteenth and early twentieth century. However, the partition of India, which also entailed the partition of Bengal into West Bengal, an Indian state, and the East Bengal, which formed the territory of Pakistan, created a great deal of havoc in the Bengali-speaking regions. The trauma of the partition and the subsequent displacement of people overshadowed the caste identity, which was slowly swiped away amid class and religious conflicts (Bandyopadhyay 2009).

Dalit lives were somehow disrupted by this sudden change of geography, identity, and nationality. Pre-partition anti-caste literary practices lost its direction completely under the current socio-political turmoil only to revive again after a fairly long period of time. The turmoil that Dalits in Bengal faced caused the absence and the subsequent delay of their literary practices to come to the forefront. Chakraborty (2021) rightly observes, "It is unfair to expect that a community that had been subjected to state-sponsored violence for decades after Independence would have the strength and ability to organise a collective cultural life and invest its energies in pursuing literary activities" (pp. 828–829). Besides, the aesthetic values it presents through literature were not accepted by mainstream Bengali writers, pushing Bengali Dalit Literature even further to the margins. Unlike Marathi Dalit Literature, it receives recognition much later. As writes Chakraborty (2021), "the (re)entry of Bengali Dalit literature into late twentieth-century Bengal did not seem to receive a favourable reception from mainstream Bengali literature, which is peculiarly dominated by writers hailing from the upper castes" (p. 820). Bengali Dalit literature flourished late compared to their Marathi and other regional counterparts due to reasons peculiar to the historical conditions of its territory.

Over a period of time, Dalit Literature has branched out into different streams in multiple Indian languages, drawing inspiration, chiefly, from Ambedkar's writings and activism in its age of infancy. Today it addresses issues that are pertinent to contemporary times. With Indian independence, the status of Dalits seems to have improved to some extent, yet newer forms of oppression have emerged. Even though untouchability has been regarded as a criminal offence, Dalits, in many cases, are still victimised by denial of access to shared ecological and geographical spaces. Their marginalisation in terms of class, caste and gender have been highlighted by scholars and activists, yet their marginal ecological circumstance demands attention. Many hues of environmental activism, and ecological studies in India have drawn attention to engendered species, threatened communities, the disastrous consequences of rapid industrialisation, commendable as well as condemnable practices in the traditional fields in agriculture, irrigation, waste-management and other allied aspects. Yet, in the words of Sharma (2012), a renowned journalist, academician and an eminent environmental critic, Indian environment studies are 'caste-blind'" by and large (p. 47). While the idea of sustainable development has been accentuated both in global and local avenues, and environmental hazards have been acknowledged as grave dangers for the entire Anthropocene, the Dalits in India are often at the frontline of this imminent predicament. Their lives, as close reading of the stories in this chapter will portray, have been relegated to the wasteland of human garbage. Yet, academic studies from the perspective of Dalit lives are too scanty to be found. Commenting on the dearth of such accounts, Sharma (2012) states that there "appears to be an environmental blindness on questions of caste, or an understanding of a specifically dalit position on ecological politics, which if taken into account can complicate our understanding of the environment, bringing to light voices of dissent and difference" (p. 46). To align this thought further, Sharma, in his *Caste and Nature: Dalits and Indian Environmental Politics* (2017) propagates the idea of "eco-casteism" that shows the justification of caste system through nature. Eco-casteism appropriates Dalits' position at an ecological fringe. The key idea here is that Dalits must be kept in conditions of constant peril that endanger their lives. This chapter is an attempt to point out how Dalits are still at the receiving ends of environmental injustices, living in denial of equal access to wealth, nature, and natural elements. Stories from *Survival and Other Stories: Bangla Dalit Fiction in Translation* (2012) edited by Sankar Prasad Singha and Indranil Acharya provides an avenue to analyse such experiences of the Dalits in Bengal.

As mentioned in the book's introduction, "This collection of translated dalit stories is primarily based in an anthology of Bangla stories published in 1996 entitled, Chaturtha Duniyar Galpo (Stories of the Fourth World)" (Singha and Acharya 2012, p. xxxiv). The stories collected here portray the lives of Dalits in Bengal, and their sufferings and discrimination in different degrees. "Survival" by Shyamal Kumar Pramanik presents Raju vis-a-vis Wild. The titular character, Munnali, is put into the dirt to survive. Sashi in "On Water and On Shore" by Gobinda Shoundo understands his perils in both the spheres. Gopal in "Farmer Gopal's Caste and Creed" shows resistance to find another job. The collection enunciates the voices of marginalised people from a *world*—distinguished by caste—within a Third World Country, India. It presents a microcosmic view of sufferings, pain, discrimination and resistance that the *fourth* world contains.

Existing scholarships on the anthology, *Survival and Other Stories: Bangla Dalit Fiction in Translation* edited by Sankar Prasad Singha and Indranil Acharya take into account the sufferings of the Dalits, focusing more on their social struggle that keep them subdued. Justin's (2016) "Under the Yoke: Representation of the Class Struggle in Select Dalit Fiction" comments on the translation of this anthology while providing analysis and observations on the heterogeneous nature of the stories: "Some stories deal with healing and redemption from the churning wheels of oppression while others deal with the pathos of being crushed under the juggernaut of superior might in the form of the upper class elite" (p. 82). The common string to hold the stories together is "the poignant rendering of previously neglected ethnic groups and the various vicissitudes in their lives" (Justin 2016, p. 83). Mehnaz Noor's "Caste in Bengal: Dalit Lives in Bengali Dalit Literature" presents a brief overview of a few stories that delineate the precarious condition in the lives of Dalits in Bengal. Those scholarships refrain from taking into account the ecological sufferings of the Dalits, suffered on a daily basis. They overlook the impact of weather change as a determining factor for the lower castes. Stories like "Munnali" bear reminiscences that Dalits are forced to live in such an ecology where dirt or filth becomes a part and parcel of their lives. Dalit as a community stands far from the prospect of environmental justice. Through an assemblage of stories with Dalit protagonists suffering in different ecological conditions, the book provides a comprehensive, albeit fictional account of relatable Dalit experiences in West Bengal.

7.2 Study of the Selected Stories

This section is a study of the four representative stories from the anthology showcasing the unique ecological plight of different kinds into which the Dalit characters are forcibly thrust:

7.2.1 *"Survival": Wresting Food from the Beast*

The first story taken for the study is titled "Survival" in English. It is titled "Ekhono Adim" in Bengali, is authored by Shyamal Kumar Pramanik, and is translated by Amrit Sen. It revolves around Raju, who for just a handful of grain, risks his life and fights with a venomous snake. He belongs to a family of landless labourers. Raju's father had come to Bardhhaman as a refugee and was helped by a local man named Bindeswar to settle in the new place. In this new found land, his family changed their occupations in different seasons: "Sometimes they wove baskets. Sometimes they worked in the field or acted as coolies at the station" (Singha and Acharya 2012, p. 144). However, during the harvest season, the family chose their hereditary occupation of picking up grains from the field. But in the chilly winter, Raju underwent the pangs of hunger as there was no job offered to him. He could manage to survive, but as a father, he could not see his hunger-stricken son. "Fatherhood seemed a heavy burden for him as tears rolled down his cheeks" (Singha and Acharya 2012, p. 144). On a chilly winter night, he ventures out to pick up the grain from a field. He comes near a rathole where he is sure that a sufficient amount of grain is stored. But getting hold of the grain is not easy as a venomous snake, cobra has wrested the place from the rat. Despite the obvious dangers to his life and many protests from his wife, Raju is hell-bent on retrieving the grain. He attempts to drive away the snake by poking the hole with a stick, the snake comes out, threatens the human transgressor and promptly returns to its sanctuary. But Raju would not let go. He pokes the snake again and this time, catches hold of the snake: "Frantically, Raju now held the snake's hood with both his hands. For a few moments time stood still as the dangerous animal and the starved man were locked in a deadly combat" (p. 145). The snake slowly accepted defeat and went away. Raju managed to feed his family by literally snatching food from the jaws of death. The food that Raju fetched was not even in human possession. Thus, his story is a typical tale of many such people for whom surviving each day comes at the cost of fighting with death.

7.2.2 *"On Water and on Shore": Treacherous Water and Treasonous Shore*

The second story that this chapter studies in the light of Environmental Casteism is "On Water and On Shore" (originally in Bengali "Jale Dangai") authored by Gobinda Shoundo, and translated to English by Subhajit Sen Gupta. The story narrates the wishful life of a lower-caste fisherman Sashi who knew "the ways of the sea" like the back of his hand (Singha and Acharya 2012, p. 65). He could even foretell the weather by gazing at the sea. Not well equipped with modern instruments, his lone guide was the star. Yet, like any mariner, he had seen the sea with its red tooth and claw. Owing to a cyclone in the middle of the ocean, his boat had once capsised leading all his fellow boatmen to seek their own safety. With courage even amidst chaos, Sashi not only managed to stay adrift, but also assisted his fellow men in staying aloft. Despite all his endeavours, he could not save his eldest son who was swept away by the current as he hopelessly watched him being lost in oblivion. Still, he continued to steer his small boat to fend for himself. But after the incident, the warehouse manager Nagen Babu suggested an inventive idea, "Tie the boat to my launch so that you may fish in the sea sitting snugly like a babu" (Singha and Archarya 2012, pp. 66–67). The idea appealed to Sashi and he did what he was told. Even as the old wounds lingered, life chose to drown itself in a fresh tide of troubles. Now the goal for Sashi was to catch as much fish as possible so that he could earn a decent income. In the sea, he moved with his boat tethered to the launch, and his spoils were carried by the lorry waiting at the shore. But even with the abundance of his catch, he was always falling short of his rewards, because the excess fish was bound to be spoilt owing to the lack of refrigeration. But soon this problem was also resolved, as Nagen Babu managed to get ice-boxes to transport the fish. One day, Sashi and his men got a boatload of fish—that too the most expensive kind *ilishes*. Sashi was overjoyed with the prospect of his rewards. His days of penury would be over, he thought. While the fish is transported safely, Nagen Babu returns the next day claiming that a heavy fine was imposed on him because the fish had spoiled due to an inadequate supply of ice. Sashi, naively accepting this story, reluctantly agrees to a meagre compensation—far less than what he truly deserved. Soon, Nagen Babu establishes his own ice factory, and Sashi is deceptively told that it was financed through a bank loan. The narrative exposes the betrayal often faced by marginalised communities, particularly those who are deprived of education and access to critical information. Sashi, who commands the waters, remains unfamiliar with the ways of the shore. His deep association with Nature grants him an intuitive understanding of Her rhythms, yet he is unable to decipher the treachery of the upper-caste Babus. Like the *ilishes,* Sashi becomes a prey for the upper castes. Sashi, like many members of his community, possesses no entrepreneurial skill and remains unaware of the ways of the world on the shore. The sea claimed the lives of his son and his fellowmen, but it was the shore that robbed him of his will to live. For their livelihood, they must brave the wild sea, yet even on land, no safety is guaranteed. Their lives are marked not only by ecological hardship but also by the social injustices rooted in caste.

7.2.3 "Munnali": The Stench of a Broken Dream

The third story in this study is titled "Munnali" in English. This story is authored by Manohar Mouli Biswas in Bengali, also named "Munnali," and is translated to English by Runa Chakraborty. The story focuses on the lives of manual scavengers and is named after its protagonist, Munnali—the teenage daughter of Kanshiram—who lives in dirt but dares to dream of a better life. She proudly declares to her friends that she will find work in the city: "I'll not stay here. I'll go to Kolkata, then cross Howrah Bridge and go straight to Mumbai by catching a huge train" (Singha and Acharya 2012, p. 100). She narrates her dream-like plan of travelling to Kolkata and then on to Mumbai with Bishu Babu, who is returning home during the Puja holidays. At first, it seems like the beginning of an adventurous journey. In reality, Munnali is hopeful for a job as a domestic help—an opportunity she believes could lift her out of the filth and despair of her current life among heaps of dirt and drains. Notably, for Munnali and others from her community, such a modest aspiration represents the highest form of dreaming. Yet, even this small hope is difficult to realise. Nathuni aunty, jealous of her prospects, comments, "Oh! What daydreams! As if these babu will take a girl from the scavenging community and keep her as a cook in their homes" (Singha and Acharya 2012, p. 100). This statement is not without reason. Munnali later learns that Bishu Babu, breaking his promise, has left town without informing anyone of his return. As a result, she is forced to remain behind and continue the age-old caste-based occupation of manual scavenging. Even the modest aspiration of working as a maid remains an unfulfilled dream for her. Instead, her mother had arranged her marriage to a cobbler considered beneath their social standing. The story ends on a bitter note: "Those who grow up in darkness look better with the ornaments of darkness" (Singha and Acharya 2012, p. 103). This symbolic ending offers a microcosmic view of the lives of manual scavengers—individuals seemingly destined to remain in filth, with no escape from their circumstances. The pervasive dirt in their lives not only crushes their dreams but also condemns them to the enduring darkness of systemic suppression.

7.2.4 "Farmer Gopal's Caste and Creed": Changing Times and Stagnant Life

The story "Farmer Gopal's Caste and Creed" (Bengali "Gopal Chashir Jatdharma") is authored by Makhanlal Pradhan, and translated in English by Goutam Buddha Sural. Gopal, the protagonist who was once allured by the sunny dreams of the budding city of Kolkata, tried his hand at various occupations—garage keeper, digger, mason—but none of them offered him any stability. Eventually, he became adamant about not trying any new job. The story contextualises the year 1972, a time when the formation of the city of Kolkata was at its dawn. It shows how lower-class people—especially farmers—suffered as government agents snatched their lands. These agents brainwashed them with false hopes of earning a living and working in the city. The loss of land left them in profound confusion. His father, however, understood the importance of agricultural land and was the last to part with it. His concern remained the same: "…if the land was gone what would they eat!" Gopal's father, Kedar Ghosh's question concerning his survival prospects meets the savoury response: "We will turn your village into a town" (Singha and Acharya 2012, p. 92). However, this prospect of a happy future soon turned to dust as the protagonist felt the dilemma regarding his occupation. Gopal, the son of Kedar Ghosh, arrived in Kolkata seeking employment. There, one could easily observe the transient nature of his affairs. As the city teemed with new automobiles, garages and motorships mushroomed in the marketplaces, Gopal secured a job at one such establishment as a handyman; however, he was neither welcomed nor given instruction in the finer arts. As new roads were built, and construction of buildings started, cheap labour was required all around the city. Gopal dug the soil in such construction venues, but that too worked only for a few days. He tried his hand at masonry too, but due to lack of enthusiasm, and being an outsider to the community of masons, he was expelled from this occupation too. Gopal, being a farmer's son, could not quite comprehend this *new* world. He thought: "In such a world what good is paddy awash in morning dew? What good does the song of the cuckoo serve in spring?" (Singha and Acharya 2012, p. 96)

It is suggestive of a world which was man-made, a land where spring lasted forever. The *new* world wiped down the villages where poor farmers lived, and it plucked them out of their ecology and space. It puts them in a dilemma, creating an identity crisis among them. Thus, after trying his hand at almost every available option, he adamantly declined to try any other profession. After the election in the 90s, Gopal found himself in a baffling position again. A small pond in his neighbourhood brought in the new idea of cultivating fish with assistance from the government. But Gopal's identity crisis became evident in his obstinacy: "He was a farmer, then he became a coolie, to a mason; and now a fisherman? He won't become one" (Singha and Acharya 2012, p. 98). The prospect of a budding city brings about certain ecological mutations that make lower caste people like Kedar and Gopal forego their primary occupation as farmers, trying out new jobs that the city has to offer. The loss of the land not only put Kedar's life in jeopardy, but stretched its effect even to Gopla's son, his grandson. Gopal's stubborn refusal to take up other forms of work put his ailing son's life at risk. The ecological shift towards urbanisation leaves Dalits caught in a constant state of uncertainty—like it disrupted his sense of identity.

7.3 Perpetuation of Untouchability

The short stories analysed in this chapter do not specifically mention any historical time frame, except the last one. Yet, a cursory reading of the tales lead to the conclusion that all of these narratives are set in independent India. Therefore, there is no explicit practice of untouchability with regard to any of the Dalit characters. All of these protagonists viz. Raju, Sashi, Munnali, and Gopal, freely mingle with people from the upper caste and seem to be fairly competent at their respective professions. Some of these characters even excel at their jobs, and therefore are appreciated by their masters and lured with better opportunities of livelihood. However, these prospects do not materialise and vanish into thin air only after creating bubbles of desire in the hearts of these lowly characters. Each story ends with an anagnorisis that the Dalits, relegated in the dirt, must seek for whatever solace that is affordable within the dirt. There is no redemption for them in the world of civilised people and they will be perpetually handed a bad deal when it comes to ecological access.

This hypothesis could be analysed with regard to the first story "Survival." Raju in this story emerges as the alpha male. In this narrative, there is a battle among representatives of three species, which are a group of rodents, a venomous snake, and a dispossessed but hungry man. The rathole originally belonged to the rodents who collected the grains and stored it in their subterranean dwelling. This space was captured without contest by the snake, which appears at a higher pedestal of the food chain. For the snake, the hole is not a storage, but a shelter where it could remain unhindered by dangers of humans or other creatures. Ideally, because of its poison, that has the capacity of incapacitating any human being, the snake is feared as well as respected by human beings. It is also revered as a religious symbol in Hinduism. That's why Matia, Raju's wife, warns him, "Are you crazy? Do you want to kill it?

Beware the Snake Goddess" (Singha and Acharya 2012, p. 143). The snake too is not an ordinary one, it is a cobra, "the king of snakes" (Singha and Acharya 2012, p. 142). Yet, Raju's act of gathering the grain as a prize by evicting forcefully the king of the snakes does not create a hopeful and victorious paeon of the strength of humanity in general. It rather paints a bleak picture of the divisive nature of human society, where Raju and his family are kept so low in the ladder of food chain. They remain still in that primordial place where the law of the jungle applies. However, it was not literally a jungle where Raju dwelled. It was an agricultural field in a village in West Bengal. As a refugee, Raju had no right over the field, nor were his services sought at the time of harvest. Thus, there are two sets of laws applicable in that field. Among the civilised human beings, Raju was an outcaste, who belonged to the periphery. He must not seek his right among humans; else he will be thrashed by the masses for any such transgression. Raju acknowledges this divide and his lower position in human society, when he says: "Damn! We can't wait! The farmers will be here soon to cut their crops. If they see us, they will rush up to thrash us. We must get the grain out. How do we get rid of the snake?" (Singha and Acharya 2012, p. 143). In the same breath, Raju is shown both as a terrified conman and as a courageous human. He eventually manages to win but only among the beasts. Among humans, he will continue to remain at the margins, begging and stealing for victuals. Thus, Raju's victory is a symbol of humanity's loss. As far as humans are concerned, he will remain an untouchable, a person who exists, yet does not exist.

There is an uncanny resemblance in the narratives of Raju's survival and Sashi's coming to terms with his position in society. Unlike Raju, Sashi is respected for his skills, his leadership, and his acumen in the aquatic sphere. The silver *illishes* that he could get in bulk are extremely expensive and are loved for their taste in Bengali culture. He had no qualms with the scary face of nature. He paid his dues by losing his son as well as his companions in the treacherous waters. But, the narrative, as was discussed above, is not about the treachery of nature, which is predictable to some extent, but about the treacherous nature of human society. His hope for better days remained futile as Nagen Sharma capitalised his labour and skills for his own advantage. Thus, Sashi and his kins remained in ignominy and penury even as the company of Nagen Babu thrived. He had to continue to serve Nagen Babu otherwise he would lose his very means of livelihood. In this story too, untouchability is not practised at the physical level. Sashi proudly recounts that Nagen Babu, a member of the upper caste, one who blew smoke of his expensive cigarette was consoled by a bidi which he shared with Sashi: "O god! To see Nagen Babu with a bidi between his lips" (Singha and Acharya 2012, p. 68). The story, therefore, justifies the hypothesis and accentuates a specific dimension of ecological marginalisation of the Dalit community. Even though natural and manmade disasters cast perils on humanity at large, the Dalits are poised at the front row. The misfortunes belong to them, while the dividends of their efforts are snatched from them and shared among the upper caste. Like Raju, Sashi too was victorious on the water. He fought with death by remaining afloat holding on to a bamboo pole and saving the lives of other as well. But among humans, he was cast to the margins. This margin too is the dangerous margin that makes him face threat to life on a daily basis. Eventually, he too experiences an anagnorisis that he, his family, and people of his community will be forever in that dark zone. Sashi "realized clearly enough that the water and the shore were not the same" (Singha and Acharya 2012, p. 70). This realisation makes him think that it would be better to throw away the excess fish as they will not be of any use to him, but of immense value to Nagen Babu and his likes.

As was discussed above, the story Munnali documents the eruption of a desire, that of a better life as a housemaid of a respectable family in Kolkata. This dream finds its nest in the heart of an innocent maiden forced to dwell in the dirt. Her "entire locality was filled with the stench of rotten filth" (Singha and Acharya 2012, p. 101). Like Raju, she too gained a few things from this abandoned wasteland and by selling them at a cheap rate. The author writes: "She along with her brothers and sisters bought some fried snacks and relished these delicacies while standing beside the newly built bypass road" (Singha and Acharya 2012, p. 102). Munnali could sense the disgust of *bhodrolok babus* while they are passing their colony, having their handkerchiefs pressing against their nose. This description makes a cogent distinction between the spaces that are occupied by the *bhodrolok* (gentry) and the spaces where people like Munnali are forced to dwell. It is difficult for people of civilised community to even move along the road without stuffing their noses with aromatic handkerchiefs. It was the same road where Munnali and her friends enjoyed the hard-earned snacks. It was the same space where they would dwell forever. Her temporary euphoria after receiving the promise that she will be taken to the town vanishes at the end. She, like her parents, siblings, and friends, will remain there, as no member of the *bhodrolok* community will accept her as a housemaid. It is noticeable that her dwelling and birth in that community also limits her occupational avenues, thus restricting her to the same occupation that her forefathers engaged in. Although her father Kansiram was not a manual scavenger, his profession in independent India was not alienated from his traditional profession. He set up latrines in the nearby areas working under the contractors who paid him a meagre sum for this arduous task. Thus, the only employment opportunity they had was to deal with filth. They were not the ones that could be hired as cooks. Like Sashi and Raju, Munnali too had this realisation at the end. While Raju was a father, and Sashi was an experienced sailor, Munnali had this realisation quite early in her life. She understood that she was, she is, and she would always remain an untouchable and thus, she should even confine her dreams according to her station in life. As Raju remained a hunter-gatherer even in the twentieth century, as Sashi remained a small-time fisherman, Munnali too would remain a scavenger. In her share will fall the realm of human garbage.

Unlike the three stories mentioned above, the fourth narrative, does provide a change in perspective. The eponymous hero does change his profession as per the need of the developing city. As was discussed above, Gopal was compelled to try his hand at different blue-collar jobs as his traditional occupation is regarded unsustainable in the fast-changing world. It was a world of motor cars, tall buildings, broad streets, and industries. If he remained steadfast in his village, he would not be able to make ends meet. At the same time, holding on to the last bastion of righteous living was no more possible for his family. They had to sell their ancestral land, as everyone else in the village had done so. There were promises of factories, offices, market places and other developments. They were, however, showered with promises of a better future: "New towns will come up in these areas. Kolkata city. Then people will get plenty of jobs. What paltry crops do you grow? Many rich people will come" (Singha and Acharya 2012, p. 92). At the same time, his change of profession does not usher any welcome change in his material condition. Due to his lower caste, he is never accepted in the society of upper caste people. As a mechanic, he was not regarded as a good apprentice and was terminated as soon as he became dispensable. The same was the case with all the other jobs where he tried to learn the trade and make progress. When he tried his hand as a mason, he faced opposition both from within and outside his community. As a lower caste Hindu, he was found an unlikely candidate to be mason and thus did not get an opportunity to learn this profitable trade as well. In Gopal's lot remained nothing, but an existential crisis at a rapidly growing material city. This realisation makes him desperately hold on to his ancestral profession as a farmer, even though it would not yield any profit for him.

7.4 Conclusion

The stories discussed above show the Dalit lives in conditions that are modern, urban, industrialised and sophisticated. Nowhere in these narratives was a single person subjected to overt caste-based slur, torture, banishment or untouchability. The fisherman Sashi shared a bidi with Nagen Babu and the latter cried on his shoulders. Farmer Gopal rubbed shoulders with people of different caste groups and religions in the burgeoning city. Munnali interacted with Bishu Babu and was offered a job in Kolkata, a city of her dreams. Raju did odd jobs in different seasons in other people's farms. None of these protagonists complained of obvious differentiated treatment meted out to them. Yet, they were all marginalised in different ways. They are the Dalits of modern India, subjected to newer forms of untouchability. They are thrown to spaces that are deemed inhospitable for the urbane masses. It is high time that along with their marginal position in society, their marginalisation in the ecological spaces be accorded due attention in academic studies.

References

Bandyopadhyay, Sekhar. 2009. Partition and the Ruptures in Dalit Identity Politics in Bengal. *Asian Studies Review* 33 (4): 455–467. https://doi.org/10.1080/10357820903363736.

Chakraborty, Runa Paunksnis. 2021. Bengali Dalit Literature and the Politics of Recognition. *South Asia: Journal of South Asian Studies* 44 (5): 817–833. https://doi.org/10.1080/00856401.2021.1962496.

Justin, Jovita. 2016. Under the Yoke: Representation of the Class Struggle in Select Dalit Fiction. *Veda's Journal of English Language and Literature (JOELL)* 3 (1). https://www.google.com/url?sa=t&source=web&rct=j&opi=89978449&url=//joell.in/wp-content/uploads/2016/01/Under-the-Yoke....pdf&ved=2ahUKEwigiY2PzJOLAxVd3DgGHXcAPDUQFnoECDIQAQ&usg=AOvVaw2FslhGYyYy0pwFg_W_bkN1. Accessed 26 January 2025.

Muktibodh, Sharatchandra. 1994. What Is Dalit Literature? In *Poisoned Bread: Translations from Modern Marathi Dalit Literature*, ed. Arjun Dangle, 267–270. Orient Longman.

Noor, Mehnaz. 2022. Caste in Bengal: Dalit Lives in Bengali Dalit Literature. Winter Issue'20. https://www.thepomelo.net/post/caste-in-bengal-dalit-lives-in-bengali-dalit-literature. Accessed 26 January 2025.

Sharma, Mukul. 2012. Dalits and Indian Environmental Politics. *Economic and Political Weekly* 47 (23): 46–52.

Sharma, Mukul. 2017. *Caste and Nature: Dalits and Indian Environmental Politics*. Oxford University Press.

Singha, Sankar Prasad, and Indranil Acharya, eds. 2012. *Survival and Other Stories: Bangla Dalit Fiction in Translation*. Orient BlackSwan.

Umesh Patra is an Assistant Professor in the Department of English at Mahatma Gandhi Central University, Motihari, Bihar. He holds a Ph.D. in English from EFLU, Hyderabad. His research area includes theatre and performance studies, modern British literature, and queer theory. A few of his lectures have been uploaded to E-Vimarsh: ICT initiative for E-learning by MGCU. Some of his writings have been published in *Economic & Political Weekly, Sanglap, Muse India*, and *The Hindu.*

Shubhadeep Jana is a Research Scholar in the Department of English, Presidency University, Kolkata. His research areas include Dalit Studies, Children's Literature, and Environmental Humanities.

Chapter 8
Stitched in Struggle: Dalit Clothing as Symbols of Environmental Struggle

Ritam Dutta and Amar Chakrabortty

Abstract The roots of Dalit environmentalism can be traced back to the historical and systematic marginalisation of Dalits in India. In medieval Gujarat, the imposition of three-sleeved shirts on Dalits emerged as a form of oppressive caste-based identification. The significance of clothing in the context of Dalit environmentalism became a tangible symbol of social hierarchy, as Dalits were compelled to wear distinct attire to ensure their visibility as lower-caste individuals. This practice not only served as a means for the upper class to easily identify and segregate Dalits but also perpetuated a degrading form of social control. The importance of clothing in this context extends beyond mere attire; it becomes a powerful tool through which the oppressive caste system manifests itself. Roland Barthes, in *The Fashion System* (2010), sees fashion as a complex system of signs. Using this theoretical framework, the research extends beyond tangible clothing items, viewing fashion as a nuanced reflection of the world that transforms from the realm of garments to the realm of rhetorical codes. Through a focused textual analysis of three Dalit poems: *The Song of Our Shirt* by Nirav Patel, *Question Paper* by M. R. Renukumar, and *Prayer* by Basudev Sunani, this chapter deliberates on the significance of clothing as a tangible symbol of social hierarchy in the context of Dalit environmentalism. Additionally, how does Roland Barthes's concept of the transition from the vestimentary to the rhetorical code provide a theoretical framework for understanding the symbolic transformation of Dalit clothing? Furthermore, in what ways can an exploration of Dalit clothing contribute to a deeper understanding of the interplay between socio-cultural, historical, and environmental dynamics in the context of Dalit environmentalism?

Keywords Clothes · Dalit · Environment · Fashion · Caste · Ecology

---

R. Dutta (✉)
Department of English, Jadavpur University, Jadavpur, India
e-mail: ritamd.english.rs@jadavpuruniversity.in

A. Chakrabortty
Department of English, Government Arts and Commerce College, Sami, Patan, Gujarat, India
e-mail: amar.chakrabortty@gujgov.edu.in

S. Kochar and R. Singh (eds.), *Pollution and Poverty*,
https://doi.org/10.1007/978-981-95-6876-5_8

8.1 Introduction

Dalit literature within the Indian literary canon dismantles dominant narratives that sustain caste hierarchies and perpetuate silences. While it clearly portrays systemic oppression, it also critically examines the socio-cultural fabric of marginalisation through the lenses of materiality and environment. Two particularly yet underexplored aspects of this discourse are Dalit clothing and ecology, which function as metaphors for resistance, identity, and survival. In caste-based societies, clothing transcends its primary role as a personal necessity or expression of individuality, becoming a site of politicised meaning. Within the Indian context, sartorial practices are deeply intertwined with questions of caste, ecology, and social justice. For Dalits, clothing symbolises both a means of oppression and a vehicle for resistance. Simultaneously, their ecological marginalisation and relegation to degraded lands and polluted spaces imbue clothing with layers of environmental and socio-political significance. This chapter, therefore, explores the intersections of Dalit clothing and ecological marginalisation, situating this analysis within the broader frameworks of caste, material culture, and eco-critical studies.

Dalit clothing exists as a marginalised element of socio-political discourse because it demonstrates substantial power in resisting oppression while asserting identity. Guru (2011), in *Humiliation: Claims and Context*, underscores the symbolic violence of sartorial restrictions imposed on Dalits. He observes that the denial of certain clothing to Dalits is not merely an act of social exclusion but an inscription of their inferior status on their bodies, ensuring that their visibility perpetually reminds others of their subjugation". Yet, Guru also highlights the transformative potential of these restrictions, arguing that the reclamation of forbidden attire becomes a radical act of defiance. The act of donning restricted garments or reclaiming forbidden attire becomes a radical assertion of equality, turning the very tools of oppression into symbols of resistance. Thus, Dalit clothing operates at the intersection of domination and defiance which functions as both an instrument of humiliation and a medium of empowerment.

Clothing, Guru (2011) insists, is an external index of humiliation under caste hegemony and simultaneously an expressive medium of defiance. Historically, caste hierarchies weaponised attire to enforce Dalits' marginalisation, relegating them to inferior and degrading dress codes. However, Dalit communities have subverted these constraints, transforming sartorial practices into acts of agency and resistance. By appropriating and reimagining clothing, Dalits challenge the aesthetic norms and socio-cultural boundaries imposed by caste oppression. These narratives gain further complexity when viewed through an ecological lens. Sharma (2017), in *Caste and Nature: Dalits and Indian Environmental Politics*, reveals the direct link between caste-based exclusion and environmental marginalisation. Sharma (2017) argues, the ecological degradation experienced by Dalits is not merely an accident of geography but a deliberate consequence of caste, wherein their exclusion from fertile lands and clean water sources relegates them to polluted and resource-scarce environments.

Historically confined to the peripheries of villages, barren wastelands, polluted industrial zones, or garbage dumps—Dalits have navigated these degraded spaces while enduring ecological and socio-political injustices. Within such contexts, clothing acquires dual significance. On one level, it reflects the environmental vulnerabilities imposed by caste, as Dalit labourers, often employed in hazardous industries like tanneries or waste management, wear garments that bear the grime of their caste-designated labour. Sharma (2017) argues that Dalit labourers wear garments that are both functional and symbolic of their socio-ecological marginalisation. These clothes, stained with the grime of caste-based labour, become markers of environmental struggle. Simultaneously, these garments embody resistance which signifies resistance amidst systemic marginalisation.

Dalit literary texts exclusively illuminate the intersections of caste, clothing, and ecological marginalisation. In the opening chapter, "Introduction: The Problem of What to Wear" of *Clothing Matters: Dress and Identity in India*, Tarlo (1996) examines the complex dimensions between dress and identity in the Indian context. Challenging the conventional academic stance that Indian identity has historically been rigidly dictated by caste and religious prescriptions, she highlights instances where individuals and communities deliberately modify or blend diverse sartorial traditions. This highlights clothing's active role in shaping both personal and collective identities. Through an extensive analysis of historical and contemporary case studies, Tarlo reveals clothing's dual function: it can simultaneously uphold and subvert existing identities, enabling the reinvention of selfhood and the adoption of new personas. Her discourse explores themes of "dress," "undress," and "re-dress," portraying sartorial choices as fluid, dynamic, and deeply embedded within India's socio-cultural fabric. By framing clothing as a site of negotiation and transformation, Tarlo critiques the orthodox view of Indian identity as unchanging, illustrating how apparel has been central to evolving notions of individuality, family, caste, region, and nation. Scholars such as Kancha Ilaiah, Gopal Guru, Chandrabhan Prasad, and Goldy George have further interrogated the casteist, nationalist, and hegemonic dimensions of environmentalism in India, often neglected by mainstream discourses. Expanding on these foundations, Sharma (2017) introduced the concept of "Dalit Ecologies," a theoretical framework that intricately maps the interrelations between caste, Dalit marginalisation, and environmental politics, providing a deeper lens into India's ecological narratives.

In *Joothan*, Vālmīki et al. (2010) recounts the indignities surrounding Dalit attire, explaining that their clothes were not our own. They came from the heaps discarded by the upper castes, worn out and tattered, yet deemed adequate for us. Even in our covering, there was humiliation. He further observes that Dalit basti [colony] was a refuse heap, filled with waste and filth. The upper castes would not even pass through it unless absolutely necessary, emphasising the material degradation imposed upon Dalits. Yet, Valmiki emphasised resilience within this oppression, by arguing each thread of those torn garments bore testimony to our survival, to our refusal to be erased despite their attempts to reduce us to nothingness. Similarly, Bama's (2012) *Karukku* critiques casteist aesthetics that regulate Dalit women's clothing, pointing towards their sartorial choices, such as wearing white saris as acts of defiance against upper-caste norms that sought to render them invisible. These narratives also reveal how caste-based spatial segregation perpetuates ecological violence, systematically denying Dalits rightful access to land and resources. The ecological dimensions of these sartorial narratives are noteworthy. Dalit labourers in tanneries and dyeing industries, as Narain et al. (2010) highlights in *Poisoned Wells: The Pollution of India's Water Resources*, endure hazardous exposure to toxic chemicals that not only harm their health but also shape their material culture. Narain argues that the clothes worn by Dalit workers in these industries are stained with the chemicals they labour amidst, turning their very bodies into archives of ecological violence. This violence is exacerbated by spatial exclusion, which relegates Dalits to polluted riversides, industrial zones, or wastelands, where even washing clothes is hindered by the lack of clean water. In *Interrogating My Chandal Life: An Autobiography of a Dalit*, the Bengali Dalit writer Byāpārī and Mukherjee (2018) explores the complex relationship between clothing and Dalit identity by also highlighting how clothing becomes a site of oppression and resistance. Byāpārī reflects on the stigmatisation of Dalits through their clothing, noting how our clothes are a mirror of our social degradation (Byāpārī and Mukherjee 2018). He contends that clothing is not merely a physical covering but a cultural marker that reinforces caste-based identities and exclusion. Yet, he also emphasises that Dalit attire, often seen as a symbol of degradation, can simultaneously be reclaimed as a form of resistance, asserting a distinct identity in the face of systemic subjugation. This complex interaction between clothing and identity illustrates the ways in which Dalits navigate both the material and symbolic dimensions of caste discrimination.

Equally critical is the ecological aspect of Dalit literature, which intertwines environmental injustice with the caste system. B. R. Ambedkar's (2014) analysis reveals how caste is not merely a division of labour but a division of labourers, enforcing an immutable hierarchy that penetrates every facet of existence, from the body to the environment. This hierarchy extends beyond social interactions to material culture, including clothing. Historically, Dalits were restricted by sartorial codes that symbolised their subjugation, often denied the right to wear new garments or any attire signifying dignity. In nineteenth-century Travancore, Dalit women were even taxed for covering their breasts, a stark example of how clothing became a tool of control. Dalits, systematically excluded from fertile land and resources, were relegated to regions of environmental degradation—wastelands and polluted areas. This spatial marginalisation was integral to the caste system, ensuring Dalits were confined to the margins of both society and ecology. As Sharma (2017) asserts, this exclusion is a structural outcome of caste, wherein the spatial and environmental vulnerability of Dalits is both deliberate and systemic, with caste creating an unequal geography, where degraded environments became synonymous with Dalit habitation.

The link between caste and ecological degradation extends beyond rural areas, permeating urban spaces where Dalits are often confined to slums or industrial zones marked by poor sanitation and environmental risks. Sharma (2017) highlights this in his study of Dalits near hazardous industries in Gujarat, observing that proximity to pollution becomes an index of caste, with the most toxic environments occupied by the most marginalised communities. Dalit literature and critical scholarship expose the environment not as neutral, but as a deeply politicised domain where caste hierarchies spatially manifest, reinforcing both ecological and social injustices. This intersectionality calls for a re-evaluation of environmental justice, advocating for the inclusion of caste in its analysis, rather than relying on simplistic binaries of nature versus human exploitation. As Sharma (2017) argues for Dalits, environmental justice is not an abstract ideal but a struggle for life itself, fought against the dual oppressions of caste and ecological degradation.

To lay the foundation for this chapter, it is crucial to incorporate Barthes' et al. (2010) "Fashion Theory." The concept of "fashion" is far from straightforward which has extended beyond mere attire to encapsulate complex socio-cultural meanings. While clothing serves a functional role in protecting the body, fashion is an abstract, universal concept. Barthes et al. (2010), in *The Fashion System*, suggests that fashion is not simply a product of clothing, but an abstract reflection of the external world. He identifies two key components: the vestimentary code, or the physical garment, and its rhetorical code, which attributes meaning to the clothing through language. This linguistic layer transforms the garment into a symbol, making fashion not just a visual or utilitarian choice, but a means of conveying a deeper statement. Barthes asserts, without discourse there is no total Fashion, no essential Fashion (Barthes et al. 2010). In this view, fashion is no longer confined to clothing, but becomes a symbol laden with cultural significance, linking language, symbolism, and society's broader cultural context. For Dalit clothing, this transition elevates practical garments into potent symbols, rich with cultural and historical weight. Barthes asserts that the true essence of fashion lies in discourse, imbuing Dalit attire with symbolic power crucial to understanding their environmental struggles. Thus, Dalit clothing becomes an emblem which is shaped by the intersection of socio-cultural, historical, and environmental forces. As Gopal Guru argues, studying Dalit clothing is to unravel the complex interplay of humiliation, labour, and ecology, where every thread tells a story of survival against the odds (2011).

8.2 The Song of Our Shirt

Despite Nirav Patel's significant influence on Dalit literature in Gujarat, Patel's work was overshadowed by the pervasive upper-caste discrimination that plagued the literary establishment. In stark contrast to his contemporary, Arun Kolatkar, whose legacy was exalted by Savarna critics, Patel found himself relegated to the margins of literary discussions. However, Patel's emergence as a transformative force in Dalit literature was not simply the result of his literary prowess; he was also an astute reader and a sensitive intellect. His approach to poetry was educational rather than confrontational. Patel's *The Song of Our Shirt* is a sophisticated exploration of caste, labour, and material deprivation, using the metaphor of clothing to expose the historical and systemic oppression endured by Dalits. Through its sharp irony and evocative imagery, the poem critiques the socio-political structures that sustain caste-based disparities, while simultaneously reclaiming agency through its subversive tone. By focusing on the shirt—not only as a tangible garment but also as a symbolic marker of caste identity, Patel crafts a layered narrative that interrogates the intersections of history, labour, and modernity. The poem highlights the Dalit experience of being compelled to wear three-sleeved shirts as a signifier of caste identity which shows the great impact of systemic discrimination on individual and cultural expression.

The poem starts by delineating a historical trajectory of suffering and dehumanisation, employing the metaphor of the shirt as its central motif. Patel further offers a trenchant critique of Dalit identity in relation to sartorial aesthetics. The declaration, "We are a fashionable caste," subverts conventional understandings of fashion by juxtaposing it with the enduring legacy of Dalit oppression, thus positioning it in stark contrast to the historical continuum of caste-based subjugation:

> "We are a fashionable caste
> Or tribe you may call:
> Our forefather Mayo Dalit
> Had a shirt of three sleeves" (Patel 2010, pp. 1–4)

The mention of the ancestral figure, Mayo Dalit, clad in a "shirt of three sleeves," serves as a resonant symbol which is deeply connected with the historical implications of clothing as an instrument of caste-based differentiation and subjugation. This metaphorical garment stands as a powerful emblem of the entrenched social hierarchy and the systemic oppression that pervades societal structures, where the number of sleeves becomes a stark marker of one's assigned caste. The lineage of oppression highlighted in the poem accentuates the inherent absurdity and brutality of caste-based hierarchies. Mayo Dalit's "three-sleeved shirt" serves as a grotesque distortion of attire, symbolising the enforced humiliation and absurdity that accompany the systematic dehumanisation of Dalits within the framework of caste orthodoxy. In this regard, Patel's portrayal links with B. R. Ambedkar's (2014) assertion that caste is not simply a social hierarchy but a pervasive system that enforces economic, cultural, and physical degradation.

Mayo Dalit's "shirt of three sleeves" aligns with Barthes' et al. (2010) notion of the rhetorical code, wherein clothing not only symbolises but transcends its physical form to function as a semiotic system, representing abstract ideas. Barthes contends, without discourse, there is no total Fashion, no essential Fashion. Interpreted through this lens, the Dalit garment transforms into a linguistic object by highlighting the complex dynamics of caste-based identity and power dynamics within the social and cultural context. The subsequent imagery of "His father had a shroud as his shirt/ And his father wore a shirt of his skin" reinforces Barthes' idea that clothing derives meaning through its naming and contextual description. In this way, the poem's shift from a shroud to a "shirt of one's skin" highlights the transformation of the physical garment into a rhetorical construct, focusing the multiple layers of symbolic meaning embedded in the attire.

Within the poem the speaker makes their declaration "I am no less fashionable" (Patel 2010) while carrying hidden defiance which generates ironic meanings. "By reclaiming the term fashionable," the speaker challenges and subverts societal conventions, destabilising established notions of fashion. The portrayal of the speaker's ragged attire, epitomised by the "pocketless, sleeveless, buttonless/Peter England, the second," emerges as a contemporary counterpoint to the historically enforced three-sleeved garments imposed on Dalits in medieval Gujarat. Nevertheless, the essential reality remains unaltered and the oppressive condition endures:

"Just got a pocketless, sleeveless, buttonless
Peter England, the second
From the pavements I sweep.
Every passerby is tempted to pay his respect" (Patel 2010, pp. 8–11)

The speaker's role as a pavement sweeper adds depth to the analysis, revealing a socio-economic reality where, despite unconventional attire, the individual performs menial, gruelling work. This juxtaposition challenges stereotypes, illustrating that clothing does not solely define one's societal rank, thus promoting a more nuanced understanding of personal expression. Patel's critique links with Guru's (2011) notion of the "politics of humiliation," where material deprivation perpetuates caste hierarchies. The speaker's "fashionable" shirt symbolises this deprivation, reflecting forced participation in a system that marginalises Dalits. By depicting the Peter England shirt as "pocketless" and "buttonless," Patel uncovers the hypocrisy of modernity, which offers Dalits not dignity but mere scraps of privilege. The line "Our shirt has a song to sing / Of bizarre fashions" encapsulates the poem's core, highlighting Roland Barthes' idea of fashion as a discourse. As Teltumbde (2010) argues, this reflects the caste economy of respect, where caste markers dictate social inclusion and exclusion. Patel's poem reveals the hypocrisy of a society that venerates material wealth while sustaining caste-based oppression.

Despite its stark imagery, *The Song of Our Shirt* transcends mere lamentation, emerging as a potent assertion of agency:

"Our shirt has a song to sing.
Of bizarre fashions," (Patel 2010, pp. 14–15)

The final lines reclaim the narrative of caste oppression, using the shirt as a symbol of resistance. Through the phrase "bizarre fashions," he compels readers to confront the absurdity of caste hierarchies, exposing the systemic inequalities that define Dalit lives. The act of singing becomes an assertion of dignity, rejecting the silence imposed by caste. This subversive tone aligns with Rege's (2013) concept of the counter-narrative of Dalit literature, which challenges the dominant caste discourse by offering alternative histories and voices. Patel's metaphorical irony thus reinforces the resilience and agency of Dalit communities in the face of systemic oppression.

8.3 Question Paper

In MR Renukumar's poem *Question Paper*, translated by M. T. Ansari, a series of probing questions about the absence, or inadequacy of specific clothing invites an exploration into the symbolic dimensions of these objects. At the outset of the poem, the imagery of birds with no doors to peck at and cots untouched by the slumberous damp of dreams suggests a deficiency in both accessibility and solace:

> "Why doesn't the house
> have doors for the birds to knock at
> and for the wind to push open?" (Renukumar 2011, pp. 1–3)

Here we could see a vivid portrayal of how the living space has been compromised with the question "why doesn't the house have doors" which reminds us of the socio-cultural and environmental conditions of the Dalits. Historically, Dalits' landlessness functioned as a key mechanism of oppression, denying them access to the agrarian wealth upon which India's socio-economic structure depended. In regions such as Tamil Nadu and Maharashtra, Dalits were often confined to the peripheries of villages, relegated to barren, arid land unfit for cultivation. This spatial marginalisation was compounded by the stigma of "untouchability," which denied them access to essential communal resources like wells, rivers, and forests. As B. R. Ambedkar (2014) critically argues in *Annihilation of Caste*, caste does not merely divide people into isolated groups; it constructs walls around resources, ensuring that Dalits are systematically deprived of life's essentials, including land and water. This ecological disenfranchisement was further exacerbated by exploitative labour practices within caste hierarchies. Dalits were often confined to "polluting" occupations, such as manual scavenging and leatherworking, which subjected them to hazardous environments. The tannery industry, for example, relied heavily on Dalit labour, while simultaneously polluting local rivers and groundwater, disproportionately affecting the communities sustaining the industry. In *Poisoned Wells: The Pollution of India's Water Resources*, Narain et al. (2010) highlights how pollution and resource scarcity are weaponised against Dalit communities, who bear the ecological costs of industries they are forced to labour in but have no share in owning.

Soon we see in the poem Renukumar emphasising on the inquiry into the incongruity between buttonless shirts and frayed shorts, which fail to align with the speaker's measurements. Renukumar writes,

> "Why are the buttonless shirts
> and the tattered shorts
> not of my size?" (Renukumar 2011, pp. 14–16)

Barthes et al. (2010) accentuates the pivotal role of discourse in the domain of fashion. Here, the dissonant attire emerges as a potent emblem which can possibly showcase the overarching societal restrictions and constraints that encumber the speaker:

> "Why do the fingers thrust

in the pocket become embarassed?" (Renukumar 2011, pp. 17–18)

In India's history of clothing, garments intended to protect from external elements have often paradoxically served as instruments of oppression, particularly for Dalits. Dr. B. R. Ambedkar's (1942) poem *Yadi Acchoot Pahanta Hai Saaf Kapdey*[1] addresses the discrimination Dalits face when donning clean clothes, poignantly questioning, "Why is there atrocity on an untouchable when he puts on clean clothes? How does that harm a Hindu?" While Gandhi used the dhoti as a symbol of resistance, Ambedkar deliberately adopted the Western-style suit as a political gesture, challenging the caste hierarchies ingrained in clothing. The suit, for Ambedkar, symbolised not only political defiance but also the dismantling of caste boundaries in a society that even dictated the dress of Dalits, derogatorily labelled as "untouchables."

Barthes et al. (2010) contends that fashion functions as a semiotic system, generating abstract concepts through its signs. Within this framework, the selection of ill-fitting and tattered clothes serves as a metaphorical articulation, mirroring not only the physical unease endured by the wearer but also alluding to a broader societal discord and disjunction between the individual and prevailing norms. Clothes devoid of buttons and worn shorts become palpable symbols of the constraints imposed upon the individual, underscoring the limitations of societal expectations. Moreover, the critique of the apparel's suitability concerning the speaker's physique invites scrutiny through Barthes' discourse-centric lens. Such mismatched attire emerges as a communicative act, emblematic of resistance against societal imperatives that demand conformity. The speaker's examination of the garment's fit signifies a subtle defiance against the rigid social structures, highlighting the alienation and discomfort faced by those who deviate from predetermined norms. Barthes' perspective affords a refined comprehension of how fashion discourse encapsulates and articulates intricate social dynamics and personal experiences.

In the end of the poem, the maternal proclamation that the speaker will unravel the answers to their queries with the advance of age introduces a temporal dimension. Roland Barthes' conceptualisation of fashion as a complex semiotic system offers a prism through which we may elucidate that these customs and rituals are deeply ingrained within a cultural matrix that may crystallise over time. This paradigm intimates an evolution in societal norms and practices, highlighting the fluidity and dynamism inherent in cultural signifiers. The temporal facet adds a layer of complexity, positing that the symbolism interwoven in these familial traditions is not immutable but rather susceptible to metamorphosis. It accentuates the complex interrelation between personal development and the shifting cultural milieu.

---

[1] Dr. B. R. Ambedkar's poem *Yadi Acchoot Pahanta Hai Saaf Kapde* ("If an Untouchable Wears Clean Clothes") addresses the discrimination Dalits face when they wear clean clothing. Ambedkar raises the critical question of why untouchables are subjected to brutality for simply dressing neatly, challenging the social conventions that perpetuate such unjust treatment. The poem highlights the deeper issue of caste-based oppression and the symbolic role of clothing in asserting dignity and defying discrimination.

8.4 Prayer

Basudev Sunani is a poet and author from Odisha, India. His poems embody a disquisition on the Dalit experience which critically interrogates the dominant perceptions held by the upper castes and inciting a reflective analysis of the deep-seated issues perpetuated by the caste hierarchy. His magnum opus, *Satyabhama*, where Sunani eschews sentimental embellishment to concentrate on the personal chronicles nested within the larger set of societal struggles. This deliberate emphasis on individual experiences forces readers to engage with the uncomfortable truths obscured by societal repression and violence. Far from succumbing to passive resignation, Sunani's versus propel the audience into a rigorous confrontation with the unsettling realities that are systematically silenced. Similarly, in his poem *Prayer*, Sunani elevates the symbolic act of disrobing to a powerful metaphor for the relinquishment of societal identities and their accompanying symbols. Roland Barthes suggested that one can unravel the multifaceted symbolism embedded in clothing and the transformative nature of disrobing.

In the beginning the poet expounds a persuasive entreaty for the abrogation of particular vestments linked to eminent personages, as illustrated by the severance of Mahapatra's coat from Brahma's physical form and the disrobing of Mohanty's suit from Vishnu's torso:

> "Come,
> Let's take off our clothes.
> Let's take off Mahapatra's coat" (Sunani and Dāsa 2008, pp. 1–3)

The poet further depicts the condition in the following lines:

> "From Brahma's body
> Let's take off Mohanty's suit
> from Vishnu's waist." (Sunani and Dāsa 2008, pp. 4–6)

The poet further depicts the situation taking the analogy of Eklavya:

> "Let's strip off Tandi's dhoti
> from Ekalavya's body
> And Satnami's langot
> from Shambuk's bottom." (Sunani and Dāsa 2008, pp. 5–10)

The discourse of advocating for the removal of particular garments embodies a desire to dismantle deeply ingrained identities and hierarchical structures within societal frameworks. This argument highlights the transformative potential that clothing, as a symbolic system, inherently possesses. The application of Barthes' rhetorical code becomes manifest as the act of disrobing undergoes a metamorphic transition, transcending its mere literal interpretation to emerge as a symbol imbued with rich cultural and historical resonance. This exhaustive inquiry goes into the subtle complexities of cultural discourse which offers penetrating insights into the complex dichotomy between attire, identity, and societal constructs. In the third stanza of the poem, the command for individuals to present themselves unclad before the divine suggests as a symbol which underscores a unified renunciation of personal vanity and societal hierarchies:

"Let us all line up
naked, in front of god
Stripped of our ego" (Sunani and Dāsa 2008, pp. 11–13)

Sunani further portrays the determination to come out of the victimization:

"we must muster our courage
for it is not easy to be naked." (Sunani and Dāsa 2008, pp. 14–15)

This examination highlights the complex nature of the struggle to liberate oneself from societal conventions and to embrace an authentic, unfiltered self. Historically, Dalits faced stringent sartorial restrictions, imposed until the reforms of B. R. Ambedkar. Their attire was subject to rigid norms, including prohibitions on garments like the dhoti, which could not fully enclose the body and had to feature pleats at the back. Restrictions also extended to items such as turbans, blouses, and the mandated black turban. In early nineteenth-century Travancore, the Nadar community was forced to expose their upper bodies as a sign of subservience, with non-compliance resulting in a punitive "breast tax." Tarlo (1996), in *Clothing Matters: Dress and Identity in India*, notes that Dalits were marked by symbols of untouchability in their clothing. Gonsalves (2012) in his book *Khadi: Gandhi's Mega Symbol of Subversion* reveals that certain professions, especially those linked to the Chandala, required the wearing of garments from the deceased which reinforce a history of suppressed autonomy in Dalits' expression through dress.

In the final lines, the poet's recognition of the divine essence of God and the deep respect for His intellectual prowess epitomizes a sophisticated interaction with the sacred. This acknowledgment highlights the potential to confront and transform entrenched societal conventions through comprehension of individual distinctiveness. The direct question posed to God, "Can you do that?" injects an element of audacious defiance and challenge against established social norms:

"Can you do that, God
Pick from the bowl
our respective surnames…" (Sunani and Dāsa 2008, pp. 30–32)

The poet continues with "and return them/ each to his own?" (Sunani and Dāsa 2008, pp. 33–34).

This inquiry suggests a ruminative exploration of the potential for liberation from societal constraints, hinting at the possibility of a reimagined identity that transcends the confines of established conventions. The act of questioning the divine to distinguish their identity without surnames suggests a strong resistance towards caste-based violence. This demonstrates how society discriminates against people on the basis of their surname and so to say with their clothes. Eco-critical theories, particularly Rob Nixon's notion of "slow violence," further illuminate this reality. Nixon (2013) observes that environmental violence is disproportionately borne by the poor and the marginalised, whose lives and livelihoods are intimately tied to the ecosystems they inhabit. In this context, clothing emerges as both a saviour to and a victim of environmental and caste-based violence which encapsulates the deep entwinement of ecological marginalisation and caste oppression. A review of existing scholarship reveals that, while the socio-political dimensions of Dalit clothing have been thoroughly investigated, its ecological implications remain insufficiently theorised.

Through this poem we could see how Renukumar's engagement engenders a more expansive dialogue that goes beyond the traditional confines of identity and conformity. The speaker's agency is manifest in their proactive interaction with the divine, eschewing mere acquiescence in favour of a deliberate navigation of and influence upon the conversation surrounding societal constructs. It highlights the orator's pivotal role in moulding and advancing a dialogue that aspires to surpass entrenched norms, thereby fostering a re-evaluation of established precepts and creating a milieu for the exploration and reconfiguration of identity.

8.5 Conclusion

Within this framework, Dalit clothing assumes a central role in understanding the community's environmental struggles, socio-cultural positioning, and relentless pursuit of justice. By approaching the garment as a symbol, one unearths its strong connection to ecological and political concerns, wherein the garment represents not just a material object but a potent site of resistance. The different types of identity, ecology, and cultural struggle within Dalit attire becomes emblematic of the ongoing fight against systemic oppression. From an ecocritical perspective, the relationship between Dalit clothing and the environment is crucial, as the physical degradation of the Dalit body and the spaces they inhabit are mirrored in the clothing they wear. As scholars like Mukul Sharma and Sunita Narain have argued, the environmental marginalisation of Dalits, relegated to the most polluted and resource-deprived zones, is closely intertwined with the material deprivation symbolised by their clothing. These garments, as both physical and rhetorical objects, symbolise not only the social and economic marginalisation of Dalit communities but also their ecological dispossession. The persistent link between caste and ecological exploitation underscores the need for a more nuanced understanding of Dalit material culture that embraces both environmental justice and social equality. In light of this, clothing becomes a crucial site for eco-critical and Dalit studies, where the act of reinterpreting and reclaiming one's attire becomes a form of resistance that contests environmental and social marginalisation. Through this analytical framework, clothing is not only a physical garment but a locus of ongoing transformation, where the boundaries of identity, environment, and social justice are constantly being redefined and contested.

References

Ambedkar, Bhimrao Ramji. 1942. *Yadi Acchoot Pahanta Hai Saaf Kapdey*. Samata. Bombay. Thacker & Co.

Ambedkar, B. R. 2014. *Annihilation of Caste: The Annotated Critical Edition*. Edited by S. Anand. With an introduction by Arundhati Roy. New Delhi: Navayana.

Bama, and Lakshmi Holmström. 2012. *Karukku*. 2nd ed. New Delhi: Oxford University Press.

Barthes, Roland, Matthew Ward, and Richard Howard. 2010. *The Fashion System*. Vintage Classics Philosophy. London: Vintage.

Byāpārī, Manorañjana, and Sipra Mukherjee. 2018. *Interrogating My Chandal Life: An Autobiography of a Dalit*. New Delhi, India: SAGE Publications India Pvt Ltd.

Gonsalves, Peter. 2012. *Khadi: Gandhi's Mega Symbol of Subversion*. Thousand Oaks, CA: SAGE Publications.

Guru, Gopal, ed. 2011. *Humiliation: Claims and Context*. Oxford India Paperbacks. New Delhi: Oxford University Press.

Ilaiah, K. 2000. *Why I Am Not a Hindu: A Sudra Critique of Hindutva Philosophy, Culture and Political Economy*. 4. repr. Calcutta: Samya.

Nixon, Rob. 2013. *Slow Violence and the Environmentalism of the Poor*. First Harvard University Press paperback edition. Cambridge, Massachusetts London, England: Harvard University Press.

Narain, Sunita et al. 2010. *Poisoned Wells: The Pollution of India's Water Resources*. Centre for Science and Environment.

Patel, Nirav. 2010. The Song of Our Shirt. *Indian Literature* 54 (1): 120–120. http://www.jstor.org/stable/23344185.

Rege, Sharmila. 2013. *Writing Caste/Writing Gender: Reading Dalit Women's Testimonies*. New Delhi: Zubaan.

Renukumar, M. R. 2011. *The Question Paper*, trans. M. T. Ansari. *No Alphabet in Sight: Tamil and Malayalam*. No Alphabet in Sight/Edited and Introduced by Susie Tharu & K. Satyanarayana; Principal Advisors, Ravikumar (Tamil), T. M. Yesudasan (Malayalam) Dossier I. New Delhi: Penguin Books.

Sainath, Palagummi. 1996. *Everybody Loves a Good Drought: Stories from India's Poorest Districts*. A Penguin Original Social Studies. New Delhi, India: Penguin Books.

Sharma, Mukul. 2017. *Caste and Nature: Dalits and Indian Environmental Politics*, 1st ed. New Delhi, India: Oxford University Press.

Sunani, Basudeva, and Jagannātha Prasāda Dāsa. 2008. *Cast Out: Poems of Anger and Angst*. 1. publ. in India. Bhubaneswar, Orissa: Rupantar.

Tarlo, Emma. 1996. *Clothing Matters: Dress and Identity in India*. Chicago, IL: University of Chicago Press.

Teltumbde, Anand. 2010. *The Persistence of Caste: The Khairlanji Murders and India's Hidden Apartheid*. 1. publ. outside of South Asia. London: Zed Books.

Vālmīki, Omaprakāśa, Arun Mukherjee, and Omaprakāśa Vālmīki. 2010. *Joothan: A Dalit's Life*. 2. repr. Kolkata: Samya.

Ritam Dutta is a doctoral research scholar in the Department of English at Jadavpur University, Kolkata. He completed his post-graduation degree from the Department of English, University of Delhi. Ritam's main areas of interest include Environmental Humanities, Victorian Literature and Marginality Studies. Ritam has coined a new literary theory called "Dalit Botanics". He is also the founder of an online forum called Victorian Vanguard which has over 20000 members worldwide. Ritam has numerous research publications through international reputed publishing houses.

Ambar Chakrabortty an Assistant Professor of English (G.E.S.Class II), is currently working at Government Arts and Commerce College, Sami in Patan district, Gujrat. He holds a Ph.D.in English from Banaras Hindu University (BHU). His research papers and book chapters have been published in various international journals and books, including Vernon Press. His areas of interest encompass Gender and Sexuality Studies, Postcolonial Literature, Queer Studies, Body Narratives, Film Studies, and ELT.

Chapter 9
Caste, Deprivation, and Ecological Disparities: An Exploration of Environmental Injustice in *Outcaste: A Memoir*

Pooja Duggal

Abstract This chapter investigates the manifestation of environmental injustice within the context of caste-based discrimination and socio-economic disparities by examining a Dalit life narrative titled *Outcaste: A Memoir* (2003) by Narendra Jadhav. In this narrative, Jadhav offers a vivid exploration of his life and the challenges faced by his family members. Through a profound analysis of the autobiographical narrative, the chapter reveals the intersectionality of caste and environmental inequality, shedding light on the systemic denial of fundamental human needs. The focus will be on exploring how Jadhav's memoir provides insights into the ways in which historically marginalised communities experience and navigate environmental challenges by unravelling connections between caste, deprivation, and ecological disparities. The chapter looks at how Dalit communities grapple with limited access to food and water. It also examines the Yeskar's role in managing death announcements and cattle carcasses, tasks that are regarded as "polluting" within the traditional Hindu social hierarchy. Through a detailed analysis of the selected text, the chapter unravels the multifaceted dimensions of ecological deprivation, tracing its roots in historical inequalities and discriminatory practices. By examining Jadhav's memoir through the lens of Environmental Casteism, this chapter not only highlights the systemic injustices faced by Dalits but also underscores the importance of integrating caste into environmental justice discourse.

Keywords Dalit · Environmental Casteism · Environmental injustice · Caste-based discrimination

P. Duggal (✉)
Department of English, Guru Nanak Dev University, Amritsar, India
e-mail: pooja.eng@gndu.ac.in

S. Kochar and R. Singh (eds.), *Pollution and Poverty*,
https://doi.org/10.1007/978-981-95-6876-5_9

9.1 Introduction

If the environment is often regarded as a common heritage of humanity, why are its benefits systematically denied to certain communities? In India, caste is not merely a social hierarchy but it also determines economic opportunities, political representation, and, crucially, access to environmental resources. Dr. Narendra Jadhav's *Outcaste: A Memoir* (2003) offers a compelling exploration of how caste-based discrimination extends beyond social and economic oppression to include environmental marginalisation. As a member of the Mahar community—historically subjected to untouchability and severe exclusion—Jadhav's lived experience provides a unique perspective for analysing how caste and ecological disparities overlap.

Dr. Narendra Jadhav, an eminent Indian economist, author, and social activist, was born in 1953. Currently serving as a Member of Parliament in the Rajya Sabha, Jadhav has had a distinguished career in public service spanning over four decades. His extensive work in economic policy, social justice, and the upliftment of marginalised communities has established him as a crucial voice in Dalit activism and policy-making. Beyond his contributions to governance, Jadhav's literary works have played a significant role in shaping the discourse on caste and social mobility in India. His *Outcaste: A Memoir* stands as a landmark in Dalit literature, providing a deeply personal yet politically charged account of his family's struggles against caste-based oppression. His family's journey—from the rural village of Ozar in Maharashtra to the urban landscape of Mumbai—mirrors the broader trajectory of the Mahar community, which historically faced severe social and environmental deprivation but sought to overcome it through education, employment, and political assertion. In addition to this seminal work, Jadhav has authored over 40 books in English, Marathi, and Hindi, including *New-Age Technology and Industrial Revolution 4.0* (2019), *Ambedkar: Awakening India's Social Conscience* (2014) and *Untouchables: My Family's Triumphant Journey Out of the Caste System in Modern India* (2005). These writings underscore Jadhav's unwavering commitment to addressing systemic inequalities in India, particularly those rooted in caste-based discrimination.

The historical roots of caste-based deprivation can be traced to India's deeply ingrained caste system, which is a rigid social hierarchy that categorises people based on birth, thereby determining their social status, occupation and access to resources. The Hindu social structure embodies a system of hierarchy and inequality, privileging the upper castes while burdening Shudras and "untouchables" disproportionately. This division, rooted in the theory of "Chaturvarna" from the Vedas, classifies society into Brahmins, Kshatriyas, Vaishyas, and Shudras, associating each with specific origins from the Purusha, forming "endogamous hereditary groups" as the social-theorist S. Gurusamy refers to them in *Dalit Empowerment in India* (2013). Beyond this hierarchy, the "Ati-Shudras" or "outcastes" occupy the lowest societal stratum due to the discriminatory practice of "untouchability." This discrimination stems from the belief that contact with lower castes can pollute or spiritually damage the higher castes.

Dalits, formerly known as "untouchables," are considered the lowest in caste hierarchy and historically faced severe social discrimination. The "untouchables" are compelled to undertake degrading tasks such as manual scavenging and face various forms of discrimination due to the perceived impurity associated with their caste. In addition to this, they face various forms of exclusion, including the denial of access to essential resources like clean water, arable land, and public spaces. Caste, therefore, is not only a marker of social exclusion but also a determinant of ecological deprivation.

The caste system has long been analysed as a mechanism of social stratification in India. Scholars such as G. S. Ghurye in *Caste and Race in India* (1932) and Dr. B. R. Ambedkar in *Annihilation of Caste* (1936) provide foundational critiques of caste-based discrimination, emphasising its role in restricting social mobility and maintaining systemic inequalities. Contemporary scholars like Suraj Yengde (*Caste Matters*, 2019) and Anand Teltumbde (*The Persistence of Caste*, 2010) extend this discussion, arguing that caste remains deeply entrenched in modern Indian society, affecting access to education, employment, and governance. While these works examine caste's social and economic dimensions, they do not extensively address its environmental implications. This gap has led to the emergence of studies exploring how caste influences ecological access and environmental justice.

Environmental justice discourse, largely developed in the context of racial discrimination in the United States, has been adapted to analyse caste-based environmental inequalities in India. Scholars such as Ramachandra Guha and Madhav Gadgil (*This Fissured Land: An Ecological History of India*, 1992) explore how India's environmental policies have historically marginalised lower-caste communities. Similarly, Amita Baviskar, in her book *In the Belly of the River* (2004), explores how development projects disproportionately displace Dalits and Adivasis, denying them access to natural resources.

Mukul Sharma, a scholar on environmental and social justice issues, builds on this scholarship by coining the term "Environmental Casteism," demonstrating how caste-based oppression extends into environmental deprivation. He astutely observes, in his book *Caste and Nature: Dalits and Indian Environmental Politics*, "caste and nature are intimately and inextricably interwoven in India; and yet their interconnectedness has rarely been a subject of examination" (2017). Similarly, the Dalit-scholar Bhimraj M. in his work "A Dalit Critique of Environmental Justice in India" draws parallels between caste and race, asserting, "'*Caste*' in India, like '*race*' in the US, is a significant factor in environmental discrimination" (2020). Environmental benefits are disproportionately withheld from individuals belonging to lower caste groups. This segment of the population faces deprivation when it comes to accessing the advantages that the environment can offer. Mukul Sharma further argues, "Nature is also an exercise in power in the hands of the powerful, and is entangled in the politics of belonging and alienation, exclusion and inclusion" (2017). He refers to this unequal distribution of environmental benefits based on caste as "Environmental Casteism," a situation where a dominant social group exercises control over environmental resources while deliberately marginalising a less privileged community, relegating them to live with the leftovers or endure the consequences of environmental scarcity.

Environmental Casteism can be seen as a form of environmental injustice, where vulnerable communities, in this case, Dalits, bear the brunt of environmental degradation and have limited resources to cope with its consequences. The concept implies a systemic and intentional pattern of environmental injustice, where upper-caste people wield disproportionate influence over resources such as land and water, while systematically excluding or limiting access for marginalised communities. Caste discrimination against Dalits can manifest in various ways, with instances extending to the outright denial of their access to essential resources, such as clean water. In addition, they face restrictions on the utilisation of communal resources, further limiting their ability to partake in communal activities or benefit from shared amenities. Furthermore, this discriminatory practice often transcends into the economic sphere, where Dalits encounter barriers in acquiring and owning land, leading to a systemic limitation on their access to one of the fundamental resources for livelihood and sustenance. In essence, these discriminatory practices create a web of challenges for Dalits, affecting not only their immediate well-being but also hindering their overall quality of life.

While scholarship on caste and environmental justice has grown, few studies integrate literary analysis with Environmental Casteism. Most discussions focus on policy and historical analysis rather than firsthand narratives. Dalit autobiographies serve as vital texts for understanding lived experiences of caste oppression and environmental deprivation. Sharankumar Limbale's *Akkarmashi* (1984) and Omprakash Valmiki's *Joothan* (1997) offer firsthand accounts of how caste-based discrimination manifests in everyday life, particularly in relation to labour and resource access. These narratives, though primarily focused on social marginalisation, implicitly reveal environmental injustices. Narendra Jadhav's *Outcaste: A Memoir* adds a crucial dimension to this discourse. This chapter seeks to bridge these gaps by analysing *Outcaste: A Memoir* as a critical text that reveals the intersection of caste, deprivation, and ecological disparities.

By examining Jadhav's memoir through the lens of Environmental Casteism, this chapter not only highlights the systemic injustices faced by Dalits but also underscores the importance of integrating caste into environmental justice discourse. Jadhav's portrayal of his family's journey, marked by resilience and struggle, provides a powerful testament to the deeply interwoven nature of caste and ecological deprivation in India. While the chapter explores broader patterns of marginalisation shaped by social hierarchies, the main focus of the analysis will remain on the denial of access of water resources and potable water. Through this analysis, the chapter aims to contribute to a broader understanding of how social hierarchies shape access to environmental resources, reinforcing historical patterns of exclusion and marginalisation.

9.2 Yeskar Duty: Exploring the Weight of Caste Obligations

Traditionally considered as "untouchables" or Dalits, the Mahars have faced social discrimination and exclusion for centuries. The Mahars are a historically marginalised community in the state of Maharashtra, India. In addition to facing discrimination, the Mahars were obligated to fulfil a duty known as "Yeskar." The Yeskar, functioning as a village servant, moved from one residence to another, delivering updates about the village, making announcements about deaths, and managing the disposal of cattle carcasses. In his memoir, Jadhav provides insight into Yeskar duty, describing it as:

> Traditionally, every Mahar household took turns performing this duty for three months. As a village servant, a Yeskar went from house to house, bearing news of the village. He announced deaths, and tended to the carcass of cattle. (2003)

This illustrates a system where individuals from the Mahar community are assigned duties that are often considered menial. This practice is indicative of historical caste-based exploitation, where certain communities were relegated to low-status occupations and tasks. The Yeskar's role in dealing with death announcements and cattle carcasses, which may be considered polluting in traditional Hindu social hierarchy, demonstrates the persistence of caste-based discrimination in social practices. He further adds:

> Any odd job that came up was assigned arbitrarily to the Yeskar, who never dared question the practice. In return, he was given a handful of grain and was allowed to go begging from house to house for leftovers. (2003)

The Yeskar's assignment of arbitrary tasks without questioning the practice implies a power imbalance. The inability to challenge the nature of assigned duties perpetuates a system where the marginalised community is subjected to exploitation without the opportunity for upward mobility or autonomy. Additionally, the compensation provided to the Yeskar, a handful of grain, and permission to beg for leftovers, indicates economic disparity and dependence. This meagre compensation further underscores the economic exploitation of the Yeskar, leaving them in a vulnerable position with limited resources for sustenance. As the Dalit ecologist Indulata Prasad, in her article "Towards Dalit Ecologies," rightly puts it, "Caste ideology not only continues to deny subjectivity and personhood to Dalits, but also dictates who is allowed to dominate the means of production, such that access to the most basic material resources is denied to Dalits" (2022). The Yeskar duty, coupled with minimal compensation, suggests a disregard for the individual's rights and a lack of consideration for their overall quality of life.

While doing his Yeskar duty, Damu was asked to guard a dead body that had surfaced in the broken well until the arrival of Fauzdar and the police party for inspection. Expressing his frustration to Namya, Damu questioned, "Why should I have to be the one to sit around without a meal since yesterday, when everybody else has his belly filled?" (Jadhav 2003). In response, Namya remarked, "Arre Damu, these are big people…how can they go hungry? That is for people like you and me. That is the way life is… just accept it, what else can we do?" (Jadhav 2003). Namya's response exposes social inequality and privilege, implying that hunger is reserved for people like Damu and himself, while the more privileged are exempt. Namya's resignation captured in the phrase "just accept it, what else can we do?" expresses a common sentiment among marginalised communities, acknowledging the systemic inequalities and power dynamics that limit their agency to challenge established norms.

As a participant in Ambedkar's social movement, Damu was reluctant to perform Yeskar duties. Sonu's observation, "I could not understand what was so bad about taking our turn as Yeskar. It was a good chance to stay in the village for a few months. After all, had not scores of generations done this duty?" prompts Damu's sharp retorts, "Can you imagine me going from house to house, knocking the ground with the jingling stick, and begging for Baluta? Can you see yourself eating the stale, half-eaten food doled out as Baluta?" (Jadhav 2003). This exchange between Sonu and Damu exhibits a dialogue about the role of a Yeskar and different perspectives on the duties associated with it. Sonu expresses a seemingly indifferent or accepting attitude towards the traditional duty of being a Yeskar, emphasising the continuity of this practice across generations.

On the other hand, Damu's retort highlights a more critical perspective, shedding light on the challenging and degrading aspects of the Yeskar's responsibilities, particularly in terms of the practice of collecting Baluta. Damu's emphasis on the degrading nature of going from house to house with a jingling stick and receiving stale, half-eaten food as Baluta, highlights the social and economic challenges faced by those performing this role. Furthermore, it's important to note that a jingling stick was employed as "it substituted the Yeskar Mahar merchant's call with the sound made by a whistle or bell, which, according to Brahminism, did not pollute caste Hindus," observed by scholars Kumar and Mishra (2022). His mention of Baluta, which refers to the meagre compensation given to the Yeskar, brings attention to the economic exploitation embedded in this practice. The act of begging for Baluta and receiving leftover or stale food reflects a form of economic disparity and the demeaning nature of the compensation associated with the Yeskar's duties. By rejecting the Yeskar duty, Damu challenges the caste system, expressing discomfort with the idea of performing tasks that are socially stigmatised or economically exploitative.

Damu further remarks, "Do you know what the upper castes say when they throw the Baluta? Take all our evils away as you take this food. Go, take the food away… better in your stomach than in the garbage, if you will take our perils with it" (Jadhav 2003). The upper castes' comments when throwing the Baluta reveal a symbolic aspect of this practice. The act of giving Baluta is not merely a form of compensation but is accompanied by a narrative that suggests transferring their burdens or evils to the person receiving it. He further adds, "How can I participate in Babasaheb's movement to reclaim dignity for our people and support these inhuman traditions?" (Jadhav 2003). Damu's rejection of inhuman traditions signifies a desire for social transformation and a break from practices that perpetuate discrimination. His concern about participating in inhuman traditions highlights the economic exploitation embedded in the Yeskar's duties. The act of taking Baluta is not just about receiving meagre food; it is a manifestation of the broader social and economic disparities that perpetuate the degradation of individuals and compromise their dignity.

Damu's internal conflict is evident in his question about supporting Babasaheb Ambedkar's movement while simultaneously being part of traditions that go against the principles of dignity and equality. This reflects the tensions faced by individuals who grapple with the need for social reform and the challenges posed by deeply entrenched cultural practices. His stance aligns with the broader goals of movements led by Ambedkar, who advocated for the annihilation of caste and the upliftment of marginalised communities. His questioning of participation in inhuman traditions indicates a recognition that individual choices contribute to the perpetuation or dismantling of oppressive social structures.

The mention of Damu's participation in Babasaheb Ambedkar's social movement signifies his support for a wider campaign advocating social justice and equal rights for marginalised communities. This newfound awareness and activism stand in contrast to his reluctance to perform Yeskar duties, which symbolise the continuation of traditional caste-based roles. Damu's resistance to participating in such duties suggests a conscious effort to break away from oppressive traditions and align himself with a movement that seeks to challenge and dismantle caste-based hierarchies. In essence, Damu's journey from a rural setting to the urban landscape of Mumbai represents not only a geographical transition but also a profound socio-cultural transformation. His experiences in the city, coupled with his involvement in Ambedkar's movement, signify a departure from the constraints of traditional caste norms and an embrace of a more assertive identity rooted in dignity, rights, and social activism.

9.3 Caste and Clean Water: Examining Water Access Disparities

In India, water emerges as a potent site for caste discrimination. The unequal distribution of water resources stands out as a fundamental issue, particularly affecting historically marginalised, lower caste communities, which encounter limited access to clean and reliable water sources. Control over water bodies often rests in the hands of dominant caste groups, enabling them to regulate access and discriminate against lower caste communities.

As Bhimraj aptly articulates, "the equitable access to natural resources is denied to the untouchables in two ways, namely, by unequal distribution of land and by imposing disability to utilize common resources" (2020). This statement is particularly relevant when considering the broader context of discrimination at water sources, where individuals of lower castes confront exclusion, humiliation, or violence. The perpetuation of traditional norms branding certain castes as "polluting" reinforces the exclusionary practices surrounding common water sources.

Consequently, members of lower castes often find themselves residing in areas with poor water quality, inadequate sanitation facilities and increased vulnerability to water scarcity—all of which result from discriminatory practices. The lack of access to clean water disproportionately impacts their livelihoods, deepening the environmental injustices they face.

Access to water, a basic necessity, remains a contentious issue for Dalits, necessitating a prolonged struggle. Unfortunately, Dalits encounter a systematic denial of equal access to this vital resource. This denial is rooted in the belief among the upper castes that the mere touch of Dalits can contaminate the purity of water, highlighting social hierarchies. The deprivation of fundamental necessities, like access to clean water, perpetuates a cycle of hardship for Dalits, exposing them to various difficulties and increasing their socio-economic vulnerabilities. The unequal distribution of water resources and control over water bodies contribute to Environmental Casteism that exacerbates existing social hierarchies.

In *Outcaste: A Memoir*, Damu, Jadhav's father, recalls an incident from his childhood where he expresses a simple desire to quench his thirst by drinking water from a large vat of water under a huge, blossoming tree. However, the vat becomes a site where caste hierarchies dictate who can directly access water. His father requests a man to let his son have a drink of water. This narrative illustrates the deep-seated discrimination and social hierarchies prevalent in the community, highlighting the impact of caste-based norms on even the most basic human needs, such as access to water.

Unaware of caste distinctions, Damu innocently extended his hand towards the tumbler. However, the man erupted in anger, exclaiming, "Son of a bitch!... How dare you try to touch this? You think you can take this from my hand?" (Jadhav 2003). This offensive reaction mirrors the prevailing societal norms that sustain discrimination based on caste identity. Damu's innocent attempt to reach for the tumbler is met with aggressive language and a derogatory remark, emphasising the ingrained belief in the perceived impurity associated with certain castes.

The refusal of the passerby to allow the narrator, a Mahar, direct contact with the water signifies a manifestation of Environmental Casteism, where the perceived impurity linked to certain castes extends to the notion of polluting the water itself. The use of derogatory language further emphasises the dehumanising treatment faced by marginalised communities in their pursuit of basic resources like water.

When Damu asked his father, "Baba, he drank water straight from the vat. Why can't I?" his father responded, "Arre, little one, we are Mahars. We cannot touch that water. If we do, we will be punished because the water will get polluted… and then no one else will be able to drink water from it" (Jadhav 2003). Baba's subsequent explanation about being Mahars and the fear of polluting the water sheds light on the social and environmental ramifications of such discriminatory beliefs. The belief that touching the water would lead to pollution and subsequent punishment, preventing others from drinking, exemplifies a deeply ingrained Environmental Casteism that determines entitlement to shared resources, while excluding certain individuals.

Furthermore, when Damu cups his hands to indirectly receive water, it becomes symbolic of the imposed hierarchy. The young narrator's innocence contrasts sharply with the harsh reality of caste-based environmental injustices, emphasising how discriminatory norms are perpetuated from one generation to the next. Even Sonu, Damu's wife, recalls one incident from her childhood related to water discrimination, saying, "Even in our village, Kherda, … we were treated as outcastes—as untouchables. I realized this early on, when I was still a child" (Jadhav 2003). There is a clear-cut exclusivist, casteist approach to the treatment of untouchables, as she further remarks:

> We had to follow strict rules regarding the river. We were never allowed to use the river water upstream like the others did, and I always wondered about it. By the way I was treated, I realized there was something wrong with me, with our entire community. (Jadhav 2003)

This incident, viewed through the lens of Environmental Casteism, accentuates how discriminatory practices are ingrained in the utilisation of natural resources. The unequal distribution of river water usage based on caste upholds a cycle of environmental injustice, where marginalised communities are denied their rightful share of essential resources. The strict rules imposed on Sonu's community regarding the usage of river water reveal a form of environmental injustice firmly established in caste distinctions. By restricting the access of the marginalised community, in this case, Sonu's family, to upstream river water, the practice propels a system where certain groups are downgraded to inferior positions and deprived of equal rights as others. The denial of access not only reinforces the notion of impurity associated with specific castes but also points to a lack of equal distribution of natural resources like water.

In *Outcase: A Memoir*, Jadhav recounts the 1927 Chavdar tank incident, where Babasaheb Ambedkar organised a meeting in Mahad to assert the rights of the "untouchables." He narrates it as:

> Here was a water reservoir from which upper-caste Hindus, and even Muslims and Christians, took water. Nevertheless, untouchables, who worshipped the same Hindu gods and who belonged to the same religion, were not allowed to take even a drop. (Jadhav 2003)

The Mahars were not allowed to drink from the Chavdar tank—"Not even when their throats were parched!" (Jadhav 2003). The incident brings attention to the denial of access to a vital resource—water—based on caste identity. While upper-caste Hindus, Muslims, and Christians freely took water from the Chavdar tank, Dalits were systematically barred from even a drop. This demonstrates a form of Environmental Casteism, where certain communities are excluded from essential resources, contributing to their vulnerability. The symbolic pollution beliefs associated with the touch of untouchables reinforce discriminatory hierarchies that extend to environmental resources. The most renowned Dalit writer and critic Sharankumar Limbale also inquires, "What is so peculiar about our touch that it pollutes water, food, houses, clothes, graveyards, tea shops, God, religion, and even man?" (2003). This raises a critical query that ought to be at the forefront of environmental inquiries in India. However, the collective assertion of rights at the Chavdar tank represents an act of environmental justice, challenging oppressive norms and reclaiming access to vital resources.

Babasaheb Ambedkar's leadership and the community's courage during this incident emerge as a form of environmental activism, contributing to a transformative narrative that aims to dismantle Environmental Casteism. Following Babasaheb's consumption of water from the tank, the narrator reflects, "We had claimed our equal rights to the water. We were equally human. Why should our touch pollute the water? We had cupped the water in our palms and sipped it! Water was still water…" (Jadhav 2003). This proclamation of equal rights to water signifies the intersectionality between environmental and social justice. In his 2021 article titled "Caste, Ecology and Nature: Environmentalism in Indian Dalit Literature," Nishanta Ghatak observes:

> In the Mahar movement, Ambedkar tried to shape the paradigm of environmental justice in India, and how it is laden with caste-issue by citing the inaccessibility of water around a village in Maharashtra. (2021)

While talking about the environmental division, Ghatak further remarks:

> [T]he environmental division across rural and urban areas is violently capitalist, binaristic and oppressive in its approach, but for the Dalits it is doubly marginalised for passing and carrying it through generations. (2021)

By challenging the discriminatory belief that their touch would pollute the water, the Dalit community not only asserts their right to a shared environmental resource but also confronts the overarching concern of environmental injustice rooted in caste-based discrimination. The act symbolises a reclaiming of agency over their environment, challenging oppressive norms that denied them access to water. The declaration emphasises the inherent quality of humanity and the universality of water, transcending caste divisions. In this context, the event becomes a powerful expression of environmental justice, as the community strives to rectify the historical environmental inequalities perpetuated by caste-based discrimination and claim their rightful place in the shared environmental commons.

However, the aftermath of the Chavdar tank event, where "the orthodox high-born of Mahad had the Brahmin priests "purify" the "desecrated" water reservoir by pouring into it 108 pots of curd, milk, cow-dung and cow-urine amidst loud religious chants!" unveils the persistence of regressive and discriminatory practices (Jadhav 2003). This ritualistic act of calling Brahmin priests to "purify" the water reservoir with substances like curd, milk, cow-dung, and cow-urine illustrates ingrained caste-based beliefs about purity and pollution. This approach intensifies the discriminatory Environmental Casteism where certain communities are deemed impure to share common resources.

Jadhav further remarks "Didn't these educated people realize that the notion of purification with animal urine was more ridiculous and despicable than the idea of defilement by human touch?" (Jadhav 2003). This remark about the absurdity of purification rituals using animal urine points to the irrational nature of these practices. This reflects not only social hierarchies but also the distortion of environmental values based on caste prejudices.

The attempt to purify the water bolsters a deeply rooted discriminatory mindset, while the Dalits' assertion of rights signifies a pivotal moment in challenging environmental injustices perpetuated by caste-based beliefs. The clash between these perspectives highlights the urgent need for dismantling discriminatory norms to achieve a just environmental landscape.

9.4 Feeding Inequality: Addressing Caste-Based Injustices in the Context of Food

Caste-based injustices have long permeated various aspects of daily life in India, intricately woven into cultural norms and societal structures. As Mukul Sharma asserts, "caste created a concept of natural and social order where people, place, occupation, and knowledge are characterized by pollution and ritual cleanliness; where bodies, behaviours, situations, and actions are isolated, 'out of place', and 'untouched', because of deep-down hierarchical boundaries" (2017). One noteworthy portrayal of this pervasive discrimination unfolds in *Outcaste: A Memoir* as Sonu, Jadhav's mother, reflects on her bewilderment as to how her touch could be deemed contaminating during a wedding feast at their landlord's house. On such occasions, they were enlisted for "odd chores such as plastering the floor with cow dung, collecting twigs and dry branches for firewood or grinding wheat on the stone wheel." At one such wedding feast, Sonu, eager to assist, took up a platter filled with sweets to distribute around. However, as soon as she started offering sweets to the guests, Heerabai, their landlord's mother, shrieked, "Eh you bloody Maharin, what do you think you are doing…You have polluted all our auspicious food, you fool. Who is going to make good for this, that one-eyed father of yours? You scoundrels, your caste will never change" (Jadhav 2003).

As Sonu extended the platter as if to give it to her, Heerabai erupted even further, exclaiming, "Now you have the audacity to touch me? Wait a minute. I will show you what it means to touch me" (Jadhav 2003). With that, she forcefully knocked the platter from Sonu's hands, causing the food to scatter in all directions. She further asserts, "Your shadow has contaminated the food. It is not fit for eating" (Jadhav 2003). Sonu stood there in confusion, pondering, "I could not figure out why the food was contaminated. Just because I had touched the platter, the food was contaminated?" (Jadhav 2003).

This incident further serves as a poignant illustration of the complex correlations between basic necessities, such as food, and discriminatory practices. Sonu, who was assigned menial tasks during the wedding feast, faces discrimination based on her caste. The discrimination extends to an absurd level where her touch is considered impure enough to make the food unfit for consumption. The mention of Sonu's confusion highlights the arbitrary nature of these discriminatory practices. Sonu could not comprehend why her touch would render the food contaminated, which emphasises the irrational and unjust aspects of caste discrimination. This incident demonstrates how caste-based biases can permeate even the most basic facets of life, affecting access to essential resources and perpetuating inequality within the realm of environmental resources. Moreover, the incident not only points to the individual suffering but also serves as an extensive critique on societal structures. By associating impurity with Sonu's caste, the discriminatory narrative expands beyond individuals to the very elements they interact with. This extension of bias, the notion that certain communities are inherently polluting, thereby exposing how even the most basic resources like food and water are not exempt from the reach of discriminatory practices.

Damu also recounts an incident from his past when he joined his friends for a wedding feast. Reflecting on the perpetual food scarcity in his household, he shares, "In our house, we always had a shortage of food. Every year something would go wrong" (Jadhav 2003). Recalling his age at around six or seven, Damu narrates how, while playing with his friends, they ventured into the grocers' lane outside the Maharwada, where the enticing aroma of sweets cooked in pure ghee captivated their senses. One of his friends, on noticing the tempting fragrance, speculated, "Hmm, looks like there's a wedding feast" (Jadhav 2003). Another friend suggested, "Let's play here…They will surely give us some leftovers" (Jadhav, 2003). However, Damu reveals, "We waited until evening, but no one offered us anything to eat, though they threw away lots of uneaten food on the garbage heap" (Jadhav 2003).

This incident once again underscores the tendency of individuals from higher castes, despite possessing ample resources, to refrain from sharing their resources with people belonging to lower castes. The episode illustrates a stark preference among the upper caste community to throw away uneaten food into the garbage rather than extending the leftovers to marginalised communities. It is also noteworthy that, in this instance, Damu and his friends were seeking "leftovers" rather than a direct share of the feast's provisions. This statement reveals a profound concern, signifying how individuals from the upper caste assign caste distinctions to resources, including those originating from the environment. This perpetuation of caste distinctions in resource allocation contributes to the pernicious influence of casteism.

9.5 Conclusion

The exploration of environmental inequality within the framework of caste-based discrimination and socio-economic disparities, as depicted in Narendra Jadhav's memoir *Outcaste: A Memoir*, elucidates the intersectionality of caste and environmental injustice. This chapter has delved into the harsh realities of Dalit lives, highlighting their restricted access to fundamental resources like food, water, and land. These revelations underscore the systemic denial of basic human needs and the deep-rooted marginalisation faced by Dalit communities.

By shedding light on the lived experiences of Dalits, this chapter contributes to a comprehensive understanding of their struggles in the face of systemic oppression. It is imperative to continue advocating for initiatives that promote equality, justice, and inclusivity, thereby striving towards a society where every individual, regardless of caste or background, can enjoy their inherent rights and dignity. This requires not only policy changes but also shifts in societal attitudes and structures, fostering genuine inclusivity and empowerment for all.

References

Ghatak, Nishanta. 2021. Caste, Ecology and Nature: Environmentalism in Indian Dalit Literature. *All About Ambedkar: A Journal on Theory and Praxis* 2: 100–112.

Gurusamy, S. 2013. *Dalit Empowerment in India*. Chennai: MJP Publishers.

Jadhav, Narendra. 2003. *Outcaste: A Memoir*. New Delhi: Viking

Kumar, Vijay, and Binod Mishra. 2022. Environmental Casteism and the Democratisation of Natural Resources: Reimagining Dalit Testimonies. *South Asia: Journal of South Asian Studies* 45: 577–595.

Limbale, Sharankumar. 2003. *The Outcaste: Akkarmashi*, trans. Santosh Bhoomkar. Delhi: Oxford University Press.

Bhimraj, M. 2020. A Dalit Critique of Environmental Justice in India. In *Contemporary Environmental Concerns: Multi-Disciplinary Aspects of Environmental Law*, ed. Renuka Soni, 89–120. Punjab: Rajiv Gandhi National University of Law.

Prasad, Indulata. 2022. Toward Dalit Ecologies. *Environment and Society* 13: 98–120.

Sharma, Mukul. 2017. *Caste and Nature: Dalits and Indian Environmental Politics*. Delhi: Oxford University Press India.

Pooja Duggal is an Assistant Professor at Guru Nanak Dev University, Amritsar. She has recently submitted her PhD thesis titled 'Caste of an Animal: A Critical Study of Selected Dalit Autobiographies' in the Department of English and Cultural Studies at Panjab University, Chandigarh. Her research areas revolve around Dalit studies and Animal studies. She has published research papers in reputed journals, such as Prose Studies, The Book Review and Dialog.

Afterword

Everyday takes the human civilization at a newer peak of scientific development and material progress. We have already sent an object beyond the solar system, mapped various terrestrial neighbours, and landed on some. The weapons of mass destruction are plentiful and the tally is increasing by the day. We are making more complex and bigger machines, and also machines to make machines. The ever-increasing application of human intelligence has created its own replication—Artificial Intelligence—to quicken the pace of 'development, growth and progress' of human civilization. Today, the part-machine part-human being is assisted by thinking machines in this venture. Thousands of satellites roaming over us, connected through electronic umbilical cords with their mother computer systems, facilitate the management of all human affairs. Creation, collection, and processing of meta-data has opened a new arena of power play because the one who has access to greater amount of data controls the system. This skewed truncated biased picture not only can be but must be redrawn to present a balanced and more 'real' picture of the current stage of the journey of the Homo sapiens—'wise man.' Speaking from one end of the spectrum of civilization today, it is possible to presume that Carl Linnaeus might have thought of some other term in its place if he were to name our species in the present era. Whatever may be the trajectory created about the affairs of the Homo sapiens, one idea prevalent in most accounts, feebly or strongly, stated or implied, could be: Man's desire to dominate his environment. If this does not appear to the case, the contemporary interpretative lens(es) would make it visible.

The world's population now exceeds 8 billion, supported by advanced medical science and technology. This progress has largely freed humanity from the fear of extinction through natural processes. However, dangers from other celestial bodies, such as asteroids or comets, still exist. These threats occasionally surface but fail to capture the attention of most people and are therefore dismissed quite easily. The fear of annihilation of the human race by its own acts, albeit, remains real. Except a nuclear war and the degradation of environment, all other potential causes are considered manageable and containable. To address the danger posed by nuclear

S. Kochar and R. Singh (eds.), *Pollution and Poverty*,
https://doi.org/10.1007/978-981-95-6876-5

weapons, several international agencies and forums have been established. Beyond that, much depends on the responsible behaviour of nuclear-armed states and other so-called 'powerful' nations. Homo sapiens are the only species that have attained power to destroy the physical world and all its inhabitants in irreparable ways. In short, man is bent upon destroying his own nest. For most of its civilizational journey, humanity carried a deep awareness of being part of nature's larger scheme. This understanding nurtured the belief in its eternal existence on Earth. That faith still exists today, though perhaps in a different form—a belief in humanity's infinite intellectual capability and, alongside it, the supposed right to subjugate the environment.

If we look at the discourse of the world culture in the past two centuries, it becomes quite obvious that this has been the most dominant aspect of the discourse and practically continues to remain so. Other voices of the discourse are usually rendered ineffective in the name of progress, development, science, technology, and even democracy. These ideas, though noble in appearance, are frequently manipulated and controlled by the dominant postulation of intellect. Binaries such as North–South, East–West, developed-developing-underdeveloped, and liberal democracy versus authoritarian regimes have all become part of the manipulative game played by contemporary, profit-driven, materialistic structures. Most accounts of Homo sapiens are shaped by this fractured discourse, which remains almost entirely under the control of the dominant belief in 'infinite intellectual capability.' Yet, voices of disagreement, challenge, and alternative visions of life do exist; though they often struggle to find space within this dominant narrative. Literary works, as part of these broader accounts, are no exception. The mechanics and dynamics of power can be traced in virtually every piece of literature.

Most loudly spoken narratives of human history in the past few centuries have brought us at the present juncture of the prime crisis of looming extinction of life on our planet. The period of a few hundred or a few thousand years is not a great chunk in the life of the development of the species. However, it is quite a long period for their extinction if the environmental factors start working against them. The telling of history is always vulnerable to manipulation. Sane, dissenting voices are routinely muted, while social and cultural institutions, practices, and values are carefully structured to mutate human groups. Such practices have been present in all civilizations which dominate the discourse today. Dalit voices are an example of the 'muted and the mutated' voices of the Indian society. But they have never been in total silence, they have always found a way to exist and persist, survive and thrive. What is also noticeable is that these muted, suppressed, and oppressed groups often become the reservoirs and carriers of the very ethos and practices that offer a ray of hope. It is through them that the possibility of saving planet Earth, and ensuring the survival of Homo sapiens from irreversible environmental degradation still exists.

Reading this book becomes a validation of the often felt and now increasingly discussed complex relationship between the physical environment and the variegated, differentiated, and hierarchically structured human groups. These chapters glean out different environmental issues embedded in the cultural discourse recreated by Dalit writers in their literary works. Since the lived experiences of these writers are rooted in the peculiar, stratified cultural system of Indian society, their narratives are both polyphonic and 'monophonic' at the same time. Together, they form a literary spectrum whose aesthetic nuances may not be fully understood or judged by mainstream aesthetic parameters—a point Dhanesh Mankulam convincingly argues in the third chapter. These Dalit writers whose works have been scrutinized in the book come from different subcultures, geographical locations, and languages of India whose gender and age group intersect with these vectors to form the cultural context of their literary works. The existential question: 'What it means to be the particular human being that s/he is?' is the main thrust of these narratives where the aesthetic parameters emanating from beyond their mundane physical, intellectual, and moral existence are either alien to them or are rejected. Most Dalit writers also maintain the truth, simplicity, and straightforwardness of the oral tradition. Hence, the classical aesthetics is less relevant. This concept is also implied in all the studies in this anthology. The focus of the researchers in this book is on laying bare the systemic processes of denial of the fruits of ecology to them on the one hand and their reverence of the environment on the other hand. The lessons that these narratives give as highlighted in the research output are: (1) environmental injustices are rooted in the cultural discourse; (2) it is possible to challenge the inflicted environmental injustice; (3) presence of ecology sensitive cultural ethos is a reality; (4) aesthetics needs to be in sync with the mechanics of the Anthropocene era; (5) narratives of the marginalized communities need to be perceived as alternate history, particularly from the ecological point of view; and so on.

Jaibir Singh Hooda

Jaibir Singh Hooda superannuated as Professor and Head, Department of English & Foreign Languages, Maharshi Dayanand University, Rohtak, Haryana. He worked in MDU, Rohtak from July 1989 to December 2022. His doctoral thesis was on African American literature but, later, he got academically inclined more towards Indian literature, folk studies, and popular literature and culture. He has taught varied courses and supervised scores of M. Phil. Dissertations and Ph. D. theses. He has been quite active in attending and organising seminars, conferences and workshops and has delivered numerous talks in these academic events. Since his superannuation he is working as Professor in the Faculty of Interdisciplinary Studies, MDU, Rohtak as Floating Faculty, and teaches Courses and conducts workshops on Communication Studies. Besides it, as Director of University Centre for Competitive Examinations, he has been organising coaching programmes for varied needs of the students of the University. He is also member of academic bodies of several universities.

GPSR Compliance
The European Union's (EU) General Product Safety Regulation (GPSR) is a set of rules that requires consumer products to be safe and our obligations to ensure this.

If you have any concerns about our products, you can contact us on

ProductSafety@springernature.com

In case Publisher is established outside the EU, the EU authorized representative is:

Springer Nature Customer Service Center GmbH
Europaplatz 3
69115 Heidelberg, Germany

www.ingramcontent.com/pod-product-compliance
Ingram Content Group UK Ltd.
Pitfield, Milton Keynes, MK11 3LW, UK
UKHW020634260926
475829UK00003BA/89
* 9 7 8 9 8 1 9 5 6 8 7 5 8 *